Moritz Paehler

Energyhomeostatic neurons of the mouse hypothalamus

AF060058

Moritz Paehler

Energyhomeostatic neurons of the mouse hypothalamus

Cellular properties of identified hypothalamic neurons that control energy balance in the mouse

Südwestdeutscher Verlag für Hochschulschriften

Imprint

Any brand names and product names mentioned in this book are subject to trademark, brand or patent protection and are trademarks or registered trademarks of their respective holders. The use of brand names, product names, common names, trade names, product descriptions etc. even without a particular marking in this work is in no way to be construed to mean that such names may be regarded as unrestricted in respect of trademark and brand protection legislation and could thus be used by anyone.

Publisher:
Südwestdeutscher Verlag für Hochschulschriften
is a trademark of
Dodo Books Indian Ocean Ltd., member of the OmniScriptum S.R.L Publishing group
str. A.Russo 15, of. 61, Chisinau-2068, Republic of Moldova Europe
Printed at: see last page
ISBN: 978-3-8381-1942-7

Zugl. / Approved by: Köln, Universität zu Köln, Diss., 2009

Copyright © Moritz Paehler
Copyright © 2011 Dodo Books Indian Ocean Ltd., member of the OmniScriptum S.R.L Publishing group

Cellular properties of identified hypothalamic neurons that control energy homeostasis in the mouse

Inaugural-Dissertation

zur Erlangung des Doktorgrades

der Mathematisch-Naturwissenschaftlichen Fakultät

der Universität zu Köln

vorgelegt von

Moritz Paehler

aus Bonn

Köln 2009

O Freunde, nicht diese Töne!
Sondern laßt uns angenehmere
anstimmen und freudenvollere.
Freude! Freude!

Ludwig van Beethoven

Berichterstatter: Prof. Dr. Peter Kloppenburg
Prof. Dr. Ansgar Büschges

Tag der mündlichen Prüfung: 02.02.2010

Contents

Abbreviations   7

Zusammenfassung   10

Abstract   13

1 Introduction   14
- 1.1 Detailed background 17
  - 1.1.1 Energy homeostasis 17
  - 1.1.2 The hypothalamic energy balance network 18
  - 1.1.3 The PVH and SIM1 neurons 21
  - 1.1.4 Importance of calcium 24
  - 1.1.5 Voltage-gated calcium channels 26
  - 1.1.6 Calcium hypothesis of brain aging 27

2 Methods   29
- 2.1 Animal care 29
- 2.2 Preparation of brain slices 30
- 2.3 Electrophysiology 31
- 2.4 Fluorimetric calcium measurements 34
  - 2.4.1 Imaging setup 34
  - 2.4.2 Experimental procedures 36
  - 2.4.3 Data analysis of fluorimetric calcium measurements 36
- 2.5 Single cell labeling and microscopy 47

3 Results — 49

3.1 POMC neuron morphology — 49
3.2 Voltage-activated calcium currents in POMC neurons — 52
3.2.1 Current/voltage relationship — 52
3.2.2 Steady-state inactivation — 61
3.2.3 Inactivation kinetics of the calcium current during a sustained pulse — 63
3.3 Calcium handling in POMC neurons in the ARC — 66
3.3.1 Calcium resting level — 66
3.3.2 Dye concentration from loading curves — 69
3.3.3 Calcium handling properties — 72
3.4 POMC and SIM1 neurons in the PVH — 77
3.5 Comparison of cellular parameters from SIM1 labeled parvocellular neurons in the PVH — 79
3.5.1 Spike frequency adaptation — 82
3.5.2 Slow afterhyperpolarization — 82
3.5.3 Tolbutamide sensitivity — 84

4 Discussion — 88

4.1 POMC neurons — 88
4.1.1 General properties — 89
4.2 Voltage-activated calcium currents in POMC neurons — 90
4.2.1 Changes of calcium current parameters — 90
4.2.2 Methodical implications — 92
4.3 Intracellular calcium handling in POMC neurons — 94
4.3.1 Changes in intracellular calcium handling — 94
4.3.2 Individual parameters — 96
4.3.3 Methodical implications — 98
4.3.4 Outlook — 99

  4.4 SIM1 neurons 101

    4.4.1 Parvocellular SIM1 subtypes in the pvPVH and mpPVH 101

List of Tables   **104**

List of Figures   **105**

References   **107**

Acknowledgements   **124**

Erklärung   **126**

Teilpublikationen   **127**

Abbreviations

$[Ca^{2+}]_i$	intracellular calcium concentration
α-MSH	α-melanocyte-stimulating hormone
γ	Ca^{2+} extrusion rate
κ_B	Ca^{2+} binding ratio
κ_S	endogenous Ca^{2+} binding ratio
τ_{endo}	endogenous decay time constant of the Ca^{2+} signal
$\tau_{transient}$	measured decay time constant of the Ca^{2+} signal
C_M	whole-cell membrane capacitance
E_M	resting membrane potential
$I_{K(Ca)}$	Ca^{2+}-dependent potassium current
K_d	dissociation constant
$K_{d,fura}$	dissociation constant of fura-2
R_s	series resistance
R_{max}	fura 2 fluorescence ratio at saturating calcium concentrations
R_{min}	fura-2 fluorescence ratio in calcium free conditions
s_{act}	slope factor of the Boltzmann equation
$V_{0.5,act}$	voltage where halfmaximal activation occurs
V_{max}	command voltage at maximal peak current amplitude
4-AP	4-aminopyridine
E_{hold}	holding potential
I_{Ca}	calcium current
aCSF	artificial cerebrospinal fluid
ADU	analog-digital units

AgRP	agouti-related protein
ARC	arcuate nucleus of the hypothalamus
ATP	adenosine triphosphate
BS	beam splitter
BST	bed nucleus of the stria terminalis
CART	cocaine amphetamine related transcript
CCD	charge-coupled device
CEA	central nucleus of the amygdala
CNQX	6-cyano-7-nitroquinoxaline-2,3-dione
D-AP5	D-2-amino-5-phosphonopentanoate
DMH	dorsomedial nucleus of the hypothalamus
EGFP	enhanced green fluorescent protein
EGTA	ethylene glycol tetraacetic acid
EM	emission
EX	excitation
FFA	free fatty acids
GABA	γ-aminobutyric acid
GaCSF	glycerol-based artificial cerebrospinal fluid
HEPES	4-(2-hydroxyethyl)-1-piperazineethane sulfonic acid
HFD	high-fat diet
HVA	high-voltage-activated
ISI	interspike interval
K_{ATP}	ATP-sensitive potassium channel
LH	lateral hypothalamic area
LPB	lateral parabrachial nucleus
LTS	low-treshold spike
LVA	low-voltage-activated

MC4R	melanocortin 4 receptor
NA	numerical aperture
ND	normal diet
NPY	neuropeptide Y
NS	parvocellular neurosecretory
NTS	nucleus tractus solitarius
PA	parvocellular pre-autonomic
POMC	pro-opiomelanocortin
PTX	picrotoxin
PVH	paraventricular nucleus of the hypothalamus
RET	reticular nucleus
ROI	region of interest
RT	room temperature
RT-PCR	reverse transcription polymerase chain reaction
SD	standard deviation
SEM	standard error of mean
SIM1	single-minded 1
TBS	TRIS-HCl buffered solution
TEA	tetraethylammonium chloride
TRIS	2-amino-2-(hydroxymethyl)-1,3-propanediol
VGCC	voltage-gated calcium channels
VMH	ventromedial nucleus of the hypothalamus

Zusammenfassung

Die Fähigkeit, Nährstoffe aufzunehmen und zu verstoffwechseln, ist überlebenswichtig für jeden Organismus. Höhere Lebewesen müssen ihre Energieaufnahme regulieren, da eine positive Energiebilanz über längere Zeit zu Fettleibigkeit und sogar zu frühzeitigem Tod führen kann. Die Energiehomöostase wird dem neurozentrischem Modell zufolge von einem kleinen neuronalen Subnetzwerk im Hypothalamus reguliert. Dieses Netzwerk umfasst, neben anderen, sattheitvermittelnde, pro-opiomelanocortin (POMC) exprimierende Neurone und hungervermittelnde, agouti-related protein (AgRP) exprimierende Neurone. Diese integrieren Signale aus der Peripherie über den Nahrungszustand des Organismus und leiten diese Informationen auf Neurone zweiter Ordnung weiter (z.B. SIM1-Neurone im paraventrikulären Nukleus des Hypothalamus; PVH). Anhand von diät-induzierten, adipösen Mäuse, die eine Diät mit hohem Fettanteil (HFD) bekommen haben, können die Effekte einer längerandauernden, positiven Energiebilanz auf dieses Netzwerk studiert werden. Im ersten Teil dieser Arbeit wurden die Effekte der HFD auf die Kalziumhomöostase von POMC-Neuronen untersucht. Kalzium spielt eine äußerst wichtige Rolle als 'second messenger' in vielen zellulären Funktionen wie z.B. der Zellmembranerregbarkeit, der synaptischen Plastizität, Neurotransmitterfreilassung und aktivitätsabhängiger Genaktivierung. Ableitungen mit der 'whole-cell patch-clamp'-Technik im 'voltage clamp'-Modus wurden an POMC-Neuronen durchgeführt, um den spannungsabhängigen Kalziumeinstrom zu charakterisieren. Außerdem wurden die Parameter für intrazelluläre Kalziumverarbeitung (Kalziumruhekonzentration, Kalizumpufferung, Kalziumextrusion) bestimmt. Dazu wurden 'whole-cell patch-clamp'-Ableitungen und schnelle optische Bildgebungsexperimente in Kombination mit dem 'added buffer'-

Ansatz verwendet. Das wichtigste Ergebnis dieser Studien war, dass eine HFD die Kalziumverarbeitung in POMC-Neuronen veränderte. Die Kalziumstromdichte war in POMC-Neuronen der Mäuse unter HFD erhöht (-69,50 pA/pF) im Vergleich zu Mäusen, die eine normale Diät (ND) erhielten (-59,63 pA/pF). Zusätzlich war der Quotient der Kalziumbindung in POMC-Neuronen der Mäuse unter HFD (220) höher als bei ND-Mäusen (399). Die Kalziumruhekonzentration war in POMC-Neuronen der HFD-Mäuse fast doppelt so hoch (0,040 µM) wie bei ND-Mäusen (0,021 µM). Diese beobachteten, diätabhängigen Veränderungen der Kalziumhomöostase könnten die Funktion, synaptische Plastizität und synaptische Leistung von POMC-Neuronen beeinträchtigen und dadurch direkt oder indirekt das Verhalten zur Energieaufnahme des Tieres beeinflussen. Zum Beispiel könnte die erhöhte Kalziumruhekonzentration durch die Aktivierung kalziumabhängiger Kaliumkanäle zu einem Verstummen der POMC-Neurone führen und dadurch möglicherweise auch zu reduzierter Vermittlung des Sattheitsgefühls. Im zweiten Teil der Arbeit wurden Kandidaten für mögliche Nachfolgeneurone zu POMC-Neuronen in einer genetisch definierten Population (SIM1-Neurone) im PVH charakterisiert. Der PVH ist eine bekannte Zielregion von POMC-Neuronen und es ist bekannt, dass SIM1-Neurone eine wichtige Rolle bei der Regulation der Energieaufnahme spielen. 'Current clamp patch-clamp'-Ableitungen in der 'whole-cell'-Konfiguration wurden an identifizierten SIM1-Neuronen durchgeführt. Subtypen von SIM1-Neuronen im PVH konnten nach ihrem Antwortverhalten auf Strominjektionen charakterisiert werden und bekannten PVH-Neurontypen zugeordnet werden. Es konnte gezeigt werden, dass SIM1-Neurone magnozellulär neurosekretorische, 'bursting', parvozellulär neurosekretorische (NS) und parvozellulär prä-autonome (PA) PVH-Neuronsubtypen umfassen. Des weiteren konnte gezeigt werden, dass die parvozellulären SIM1-Subtypen unterschiedlich sensitiv für den selektiven Blocker des adenosintriphosphat-sensitiven Kaliumkanal (K_{ATP}) Tolbutamid sind (22.23 % der abgeleiteten NS-Neuronen reagierten auf Tolbutamid und 92,31 % der PA-Neurone). Dies könnte auf unterschiedliche Expression des K_{ATP}-Kanals innerhalb der jeweiligen SIM1-Subtypen und damit auf die Existenz verschiedener NS und PA-Subpopulationen hindeuten.

Abstract

The ability to take up and metabolize nutrients is vitally important for the survival of every living organism. Higher organisms need to control their energy uptake and expenditure as a positive energy balance over time can lead to obesity and even early death. The regulation of energy homeostasis, according to the neurocentric model, is performed by a small neuronal subnetwork in the hypothalamus. This network comprises, among others, satiety mediating, pro-opiomelanocortin (POMC) expressing neurons and hunger mediating, agouti-related protein (AgRP) expressing neurons. They integrate signals from the periphery about the nutritional status of the organism and relay this information onto second order neurons (e.g. SIM1 neurons in the paraventricular nucleus of the hypothalamus; PVH). Diet-induced obese mice that have received a high-fat diet can be used to investigate the effects of a prolonged positive energy balance on this network. In the first part of this thesis, the effects of a high-fat diet on the calcium homeostasis of POMC neurons were investigated. Calcium plays a crucial role as a second messenger in many important cellular functions such as membrane excitability, synaptic plasticity, neurotransmitter release and activity dependent gene activation. Whole-cell patch-clamp recordings in voltage clamp were performed to characterize voltage-activated calcium influx in POMC neurons. Further, to determine the intracellular calcium handling parameters (calcium resting level, calcium buffering, calcium extrusion) whole-cell patch-clamp recordings and fast optical imaging were used in combination with the 'added buffer' approach. The most important finding was that the high-fat diet affected calcium handling in POMC neurons. The calcium current density was found to be increased in POMC neurons of high-fat diet mice (-69.50 pA/pF) compared to normal diet mice (-59.63 pA/pF). Furthermore, the calcium binding ratio of POMC neurons in the high-fat diet cohort was lower (220) compared to normal diet mice (399). Also, the calcium resting level of POMC neurons was almost twice as high in the high-fat diet cohort (0.040 µM) as in the normal diet mice (0.021 µM). These observed diet dependent changes in calcium handling could impair the function, synaptic plasticity and synaptic output of POMC neurons, thus directly or indirectly influencing the energy uptake behaviour of the animal. For example, the increase in the calcium resting level could lead to a silencing of POMC neurons from high-fat diet mice through the activation of calcium dependent potassium channels, possibly reducing satiety signaling. In the second part of this thesis, candidates for putative second order neurons to POMC neurons in a genetically defined neuron population in the PVH (SIM1 neurons) were characterized. The PVH is a known target region of POMC neuron terminals and it was shown by earlier studies that SIM1 neurons in the PVH play an important role in energy uptake regulation. Current clamp patch-clamp recordings in the whole-cell configuration were performed on identified SIM1 neurons. Subtypes of SIM1 neurons in the PVH could be characterized according to their reaction to current injection protocols and matched with neurontypes found in previous studies. It could be shown that the SIM1 neuron population comprise the magnocellular neurosecretory, bursting, parvocellular neurosecretory (NS) and parvocellular pre-autonomic (PA) PVH neuron subtypes. Additionally, it was found that the parvocellular SIM1 neuron subtypes differ in their sensitivity to the adenosine triphosphate-sensitive potassium (K_{ATP}) channel blocker tolbutamide (Only 22.23 % of the recorded NS subtype neurons were sensitive to tolbutamide and 92.31 % of the PA subtype) suggesting a possible differential expression of the K_{ATP} channel for each subtype and the possible existence of subpopulations of NS and PA neurons.

1 Introduction

Every living organism needs to take up and metabolize nutrients. This behaviour needs to be tightly regulated. In mammals, the hypothalamus regulates energy uptake and expenditure by integrating afferent and humoral input that conveys information about the nutritional state of the body (Schwartz and Porte, 2005). Disregulation of this system can lead to obesity and the metabolic syndrome including type 2 diabetes and heart disease (Huxley *et al.*, 2009). These are a major health risks and a great public health threat to western societies (World Health Organization, 2009; Finkelstein *et al.*, 2009). The current, neurocentric network model of energy homeostasis proposes that antagonistic first-order neurons in the arcuate nucleus of the hypothalamus (ARC), such as pro-opiomelanocortin (POMC) neurons and agouti-related protein (AgRP) neurons, integrate information coming from, among others, peripheral tissues (e.g. adipose tissue, pancreas, gut). These neurons synapse onto second-order neurons (possibly SIM1 neurons; Balthasar *et al.*, 2005) in many brain regions and subnuclei of the hypothalamus, e.g. the paraventricular nucleus of the hypothalamus (PVH; Swanson, 2000; Cone, 2005; Fig. 1.1).

In many neuronal processes, calcium plays an important role, such as synaptic plasticity, release of neurotransmitter, membrane exictability, enzyme activation and activity dependent gene activation. Its cytosolic concentration is elevated mainly by influx through voltage-gated Ca^{2+} channels (VGCC). Because of its many functions, Ca^{2+} signaling has to be tightly regulated in duration and location inside the neuron (Augustine et al., 2003). Disregulation of Ca^{2+} homeostasis can lead to impairment of neuronal function, synaptic activity and synaptic plasticity. For example, subtle changes in neuronal Ca^{2+} homeostasis seem to be a major factor for the cognitive decline in old age. Interestingly, this can be attenuated by caloric restriction (Toescu and Verkhratsky, 2007; Murchison and Griffith, 2007).

To better understand hypothalamic energy balance, diet-induced obese (Archer and Mercer, 2007) and knockout mice (Sauer and Henderson, 1988; Gu et al., 1994) can be generated and studied by analyzing the performance of the animal in various tests. However, comparatively little knowledge is gained about the properties of the neurons within the hypothalamus and their cellular parameters. The difficulty of identifying these neurons in the living tissue has hindered the study of single neurons with whole-cell patch-clamp electrophysiology within this network for a long time. Recent advances in genetics now allow to label specific neurons in the hypothalamus by expressing fluorescent proteins, such as enhanced green fluorescent protein (EGFP), under the control of neuron specific promoters. The mice used in this thesis express EGFP under the transcriptional control of mouse *Pomc* and *Sim1* genomic sequences (Fig. 1.2, C & D; Fig. 3.12, B, C; Cowley et al., 2001; Balthasar et al., 2005). This allows to distinguish POMC and SIM1 neurons unequivocally from other cells in the living tissue by their EGFP fluorescence and perform whole-cell patch-clamp recordings from identified neurons. This further allows to investigate the cellular properties (e.g. neuroanatomy, active membrane properties, Ca^{2+} handling) of genetically identified neurons in the energy homeostatic network. In addition, these data can be combined with insights gained from neuron specific knockout mice, joining two very

1 Introduction

powerful analytical toolsets. Ultimately, this will help to gain a better understanding of energy homeostasis, its principal components and their properties that underlie the regulation of food intake and energy expenditure. This understanding will enable us to come up with novel solutions for the social and economic burden of obesity and obesity-related diseases.

Aim of this thesis

The aim of this thesis was to better comprehend the components (and their cellular properties) of the hypothalamic network that regulates energy balance and establish a framework for future studies. The goal was approached in two parts.

First, to better understand calcium handling in first order neurons of the energy homeostatic network, (a) the calcium influx in POMC neurons was characterized by performing whole-cell patch-clamp voltage clamp recordings in hypothalamic brain slices, from mice fed either a high fat or a normal diet, and (b) the calcium buffering capacity and extrusion rate of POMC neurons of these mice were determined using the 'added buffer' approach. This was achieved by performing ratiometric calcium imaging experiments in combination with whole-cell patch-clamp recordings of identified POMC neurons in hypothalamic brain slices.

Second, potential candidates for second-order neurons to POMC neurons in the paraventricular nucleus of the hypothalamus were characterized performing whole-cell patch-clamp current clamp recordings in a genetically identified subpopulation of PVH neurons: SIM1 neurons.

1 Introduction

1.1 Detailed background

In order to survive, every living organism needs to take up and metabolize nutrients. In vertebrates this behaviour needs to be tightly regulated because even a minute imbalance in energy homeostasis can lead to obesity, disease and even early death. In humans, being obese is a major risk factor for hypertension, stroke, cancer and metabolic syndrome including type 2 diabetes and heart disease (Must *et al.*, 1999; Huxley *et al.*, 2009; Osmond *et al.*, 2009). Currently, the increasing prevalence of obesity and obesity-related diseases in the western society are a great public health threat of pandemic proportions (World Health Organization, 2009; Finkelstein *et al.*, 2009). As obese people grow older, the risk of manifestation of these diseases increases, especially in the second half of life (Thompson *et al.*, 1999). Obesity also increases the risk of reduced cognitive function, accelerates cognitive aging and increases the risk neurodegenerative disease, such as dementia and Alzheimer's disease (Luchsinger and Mayeux, 2007; Yaffe, 2007). In rodents, obesity can be elicited by feeding a high fat diet (diet-induced obesity) because it mimics the unnatural environment of modern society with minimal essential activity and access to highly-nutritious, energy-dense food (Surwit *et al.*, 1988; West *et al.*, 1992; Archer and Mercer, 2007). A better understanding of energy homeostasis and its principal components will help to devise strategies to overcome this social and economic scourge.

1.1.1 Energy homeostasis

In healthy vertebrates, the brain regulates food intake and energy expenditure in response to the body's nutritional status (Figure 1.1). The nutritional status is conveyed by peripheral signals of the pancreas, adipose tissue and the intestinal tract. These

1 Introduction

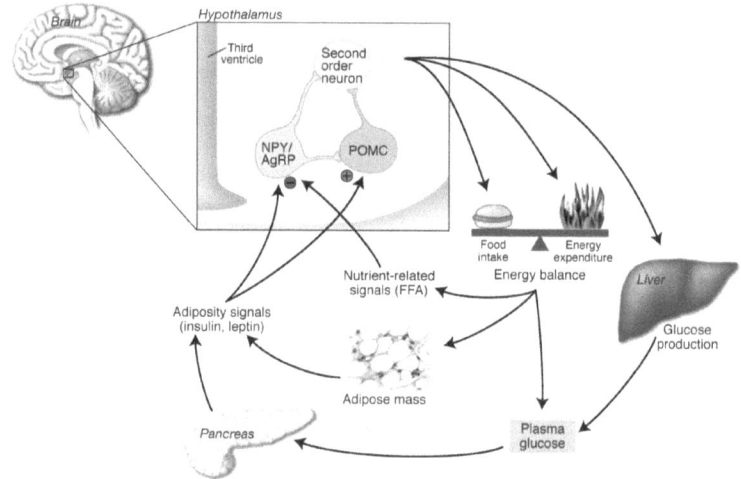

Figure 1.1: Diagram depicting the neurocentric model of energy homeostasis. Food intake and energy expenditure are regulated by the brain in response to afferent input and signals from peripheral tissue. These signals include hormones, such as insulin and leptin, but also nutrient related signals, such as free fatty acids and glucose. A small neuronal subnetwork in the hypothalamus containing AgRP/NPY and POMC neurons integrates signals about the nutrient state of the organism and regulates substrate metabolism and food intake accordingly. AgRP: Agouti-related protein, FFA: free fatty acids, NPY: neuropeptide Y, POMC: pro-opiomelanocortin. Modified from Schwartz (2005).

signals are released into the blood and reach the brain together with afferent input (Arora and Anubhuti, 2006; Morton et al., 2006). Among others, two major signals are identified today: leptin (produced by adipose tissue) and insulin (produced by the pancreas). Both peptides are released into the blood proportionally to the amount of body fat and pass the blood-brain barrier in proportion to their plasma concentration (Saltiel and Kahn, 2001).

1.1.2 The hypothalamic energy balance network

Lesion studies in the 1940s showed that there are discrete regions of the hypothalamus that are essential in regulating food intake (Fig. 1.2, A & B). Destruction of the hy-

1 Introduction

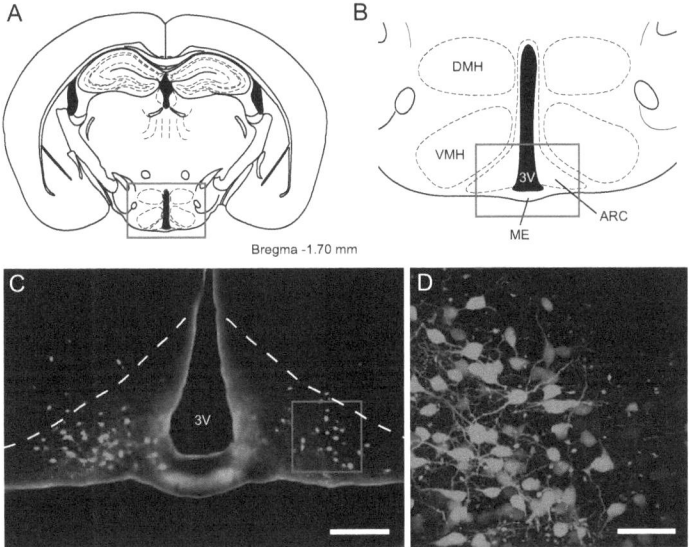

Figure 1.2: Overview of hypothalamic nuclei involved in energy homeostasis and the location of POMC neurons expressing enhanced green fluorescent protein (EGFP). **(A)** Schematic representation of a coronal section of the mouse brain. The area of the red box is shown in B. **(B)** Overview of hypothalamic nuclei involved in energy homeostasis including the dorsomedial, ventromedial and arcuate nucleus of the hypothalamus. The area of the red box is shown in C. **(C)** Merged fluorescence and transmission image depicting the arcuate nucleus (outlined by the dashed line) and POMC neurons expressing EGFP. The area of the red box is shown in D. **(D)** Maximum intensity projection of a fluorescence image stack of POMC neurons expressing EGFP in the arcuate nucleus. Fluorescence images were acquired in the living slice using multiphoton microscopy. Scale bar in C: 200 µm, in D: 50 µm. 3V: third ventricle, ARC: arcuate nucleus of the hypothalamus, DMH: dorsomedial nucleus of the hypothalamus, ME: median eminence, VMH: ventromedial nucleus of the hypothalamus. A and B modified from Paxinos and Franklin (1997).

pothalamic arcuate (ARC), ventromedial (VMH), dorsomedial (DMH) and paraventricular (PVH) nuclei in rats resulted in severely obese animals. Contrarily, destruction of the lateral hypothalamic area (LH) lead to a decrease in food intake (Hetherington and Ranson, 1940; Miller *et al.*, 1950; Stevenson, 1970).

In the mid-nineties, the discovery of leptin and the gene encoding it, by means of obese, leptin deficient *ob/ob* mice, marked another great step in the understanding of energy regulation (Zhang *et al.*, 1994; Halaas *et al.*, 1995; Pelleymounter *et al.*, 1995;

Campfield *et al.*, 1995). Additionally, leptin receptors could be cloned and localized in the ARC and the VMH (Tartaglia *et al.*, 1995; Mercer *et al.*, 1996). Apart from classical insulin target regions (in skeletal muscle, adipose tissue and the liver), insulin receptors are also found in the brain where specific insulin receptor knockout resulted in obese mice (Brüning *et al.*, 2000). Injection of insulin into the cerebral ventricles of mice resulted in increased expression of the anorectic peptide precursor protein pro-opiomelanocortin (POMC) and reduction of food uptake (Benoit *et al.*, 2002).

To this day, a model has emerged which involves two antagonistic types of hypothalamic neurons located in the ARC: neurons expressing the satiety mediating (orexigenic) polypeptides pro-opiomelanocortin (POMC) and cocaine amphetamine regulated transcript (CART), and neurons which express the hunger mediating (anorexigenic) peptides agouti-related protein (AgRP) and neuropeptide Y (NPY; Hahn *et al.*, 1998; Cone, 2005). In the current model, POMC neurons are activated by leptin, triggering the release of α-melanocyte-stimulating hormone (α-MSH) from POMC axon terminals. α-MSH in turn activates the melanocortin 4 receptor (MC4R) in second-order neurons (e.g. in the PVH) leading to reduced food intake and increased energy expenditure (Elias *et al.*, 1999; Cowley *et al.*, 2001). Secondly, leptin suppresses NPY/AgRP neuron activity, which in turn would antagonize the activity of α-MSH. Furthermore, NPY/AgRP neurons also inhibit POMC activity through the action of NPY and γ-aminobutyric acid (GABA) releasing synapses onto POMC neurons leading to increased food intake.

POMC neurons from the ARC project to a diverse range of locations in the brain. These target areas include: the bed nucleus of the stria terminalis (BST), the central nucleus of the amygdala (CEA), the lateral parabrachial nucleus (LPB), the lateral hypothalamic area (LH), the nucleus tractus solitarius (NTS), the reticular nucleus

(RET) and the paraventricular nucleus of the hypothalamus (PVH; reviewed in Cone, 2005).

1.1.3 The PVH and SIM1 neurons

In addition to controlling ingestive behaviour of food and drink, the PVH is also thought to be involved in neuroendocrine and autonomous regulation of the cardiovascular system, water homeostasis, milk ejection and uterine contraction during child birth (Malpas and Coote, 1994; Stocker *et al.*, 2004; Lincoln and Wakerley, 1974; Summerlee and Lincoln, 1981). The PVH is a complex hypothalamic nucleus that can be divided into at least eight, clearly distinguishable subdivisions and contains approximately 30 different putative neurotransmitters (Swanson and Sawchenko, 1983). The cellular parameters of PVH neurons have been studied previously in great detail and several neuron types could be characterized (Tasker and Dudek, 1991; Hoffman *et al.*, 1991; Tasker and Dudek, 1993; Luther and Tasker, 2000; Luther *et al.*, 2000; Stern, 2001; Luther *et al.*, 2002; Melnick *et al.*, 2007). In these studies, the neuron types were found to be confined to relatively discrete subdivisions of the PVH and could be classified not only by their location but also by soma size, immunoreactivity and electrophysiological response to current injection. The following classes of neuron types have been identified so far and can be grouped into three main categories: (1) neurosecretory magnocellular, (2) bursting and (3) parvocellular neurons. Magnocellular neurons are located in the ventral and lateral subdivisions of the PVH and project to the neurohypophysis. Bursting neurons are located laterally and ventrolaterally of the PVH and the third ventricle. They are hypothesized to be inhibitory interneurons (Tasker and Dudek, 1991). Parvocellular neurons are generally located in the more medial and dorsal regions of the PVH. They can be subdivided into (a) parvocellular neurosecre-

1 Introduction

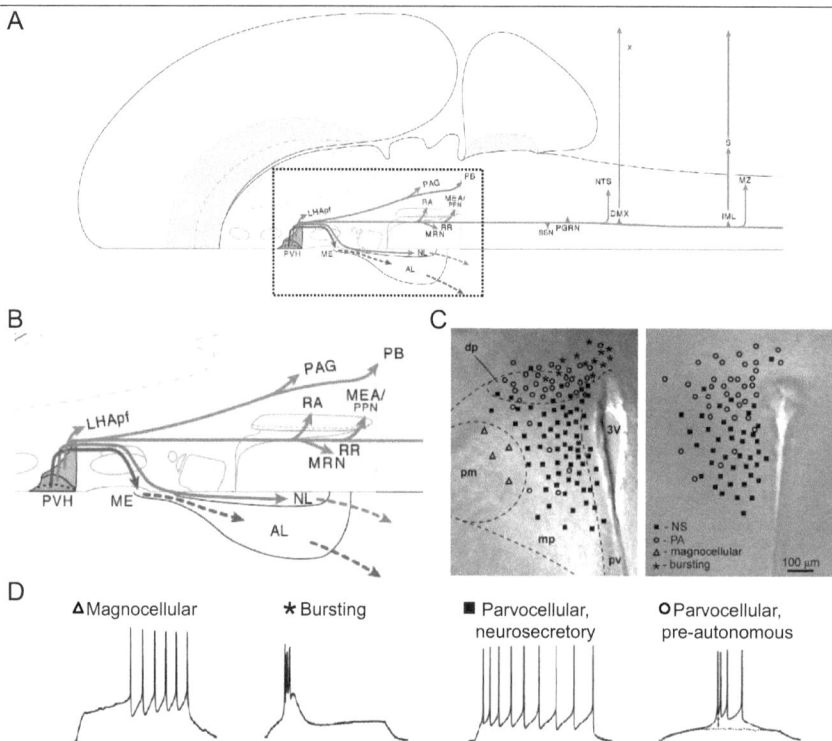

Figure 1.3: Output projections from the PVH and the corresponding neuron types. (A) Schematic representation of a dorsal view on a mammalian brain hemisphere illustrating the output projections originating from the PVH. (B) Enlarged view of area marked with a dotted line in A of the short range projections originating from the PVH. The projections of magnocellular neurons are shown in blue, of parvocellular neurosecretory neurons in red and those of parvocellular pre-autonomous neurons are shown in green. (C) Schematic overlay of a transmission image of the PVH visualizing the distribution of the different neuron types. (D) Electrophysiological profiles of PVH neurons. Representative traces of the membrane potential during positive current injection. Magnocellular neurons are characterized by a delayed onset in action potential firing upon depolarization. Bursting neurons display large, high-frequency bursts of action potentials superimposed on low-threshold spikes during depolarization. Parvocellular neurosecretory neurons display tonic action potential firing and weak spike-frequency adaptation. Parvocellular pre-autonomic neurons fire a low-threshold spike upon hyperpolarization or by depolarization from membrane potentials more hyperpolarized than the resting membrane potential. 3V: third ventricle, AL: anterior lobe pituitary, dp: dorsal cap, DMX: dorsal motor nucleus vagus, IML: intermediolateral preganglionic column, LHApf: perifornical lateral hypothalamic area (tuberal level), mp: medial parvocellular subdivision, ME: median eminence, MEA: midbrain extrapyramidal area, MRN: mesencephalic reticular nucleus, MZ: marginal zone, NL: neural (posterior) lobe pituitary, NTS: nucleus of the solitary tract, PAG: periaqueductal gray, pm: posterior magnocellular subdivision, pv: periventricular parvocellular subdivision, PB: parabrachial nucleus, PGRN: paragigantocellular reticular nucleus (ventrolateral medulla), PPN: pedunculopontine nucleus, RA: raphe nuclei, RR: retrorubral area, S: sympathetic ganglia, SSN: superior salivatory nucleus, X: vagus nerve. A and B modified from Swanson (2000). C and D were modified from Melnick et al. (2007).

tory neurons that project to the median eminence regulating hormone release from the anterior pituitary gland and (b) parvocellular pre-autonomous neurons (PA) that send long projections to the autonomic brain centers in the brain stem and spinal cord (Swanson *et al.*, 1980; Sawchenko and Swanson, 1982; Swanson and Sawchenko, 1983; Swanson, 2000, Fig. 1.3, A – C). The classification of these neuron types correlates with their electrophysiological responses to current injection (Fig. 1.3, D). Magnocellular neurons are characterized by a delayed onset in action potential firing upon depolarization and show weak spike-frequency adaptation. Bursting neurons display large, high-frequency bursts of action potentials superimposed on sometimes repetitive low-threshold spikes during depolarization (Tasker and Dudek, 1991). Parvocellular neurosecretory neurons respond to depolarizing current injection with tonic action potential firing and weak spike-frequency adaptation (Luther *et al.*, 2002). Parvocellular pre-autonomic neurons fire a low-threshold spike upon hyperpolarization or by depolarization from membrane potentials more hyperpolarized than the resting membrane potential (Tasker and Dudek, 1991).

The periventricular (pv) and medial parvocellular (mp) subdivisions of the PVH receive a dense innervation from POMC neuron fibers. The pv- and mpPVH are also regions with a high density of α-MSH receptor (MC4R) expression, suggesting that these are target regions where POMC neurons synapse onto second-order neurons (Elmquist *et al.*, 1997; Bagnol *et al.*, 1999; Cowley *et al.*, 1999; Bouret *et al.*, 2004). Additionally, Balthasar and colleagues were able to show that hyperphagia induced by the knockout of the MC4R could be rescued by selective restoration of MC4R in neurons expressing SIM1 in the PVH. This finding suggests that there is (a) a divergence in the pathway regulating energy uptake and energy expenditure and, more importantly, (b) that SIM1 neurons in the PVH play an important role in this pathway (Balthasar *et al.*, 2005).

SIM1 was originally identified in the fruit fly *Drosophila* as part of the single-minded gene family where it plays a critical role in the development of the midline neuroepithelium during embryonic development (Nambu *et al.*, 1991; Fan *et al.*, 1997). Evidence also suggests that SIM1 is needed for proper axon targeting during the development of the mammalian CNS (Marion *et al.*, 2005). In humans, haploinsufficiency of the SIM1 homolog is associated with hyperphagic obesity and Down syndrome (Fan *et al.*, 1997; Holder *et al.*, 2000; Faivre *et al.*, 2002). In mice, SIM1 heterozygous animals display hyperphagia, increased linear growth and susceptibility to diet-induced obesity (Ema *et al.*, 1996).

Despite this amount of information, it is still unclear what the electrophysiological properties of SIM1 neurons are and which known PVH subtypes they contain.

1.1.4 Importance of calcium

Calcium plays a crucial role in many important cellular functions. In neurons, Ca^{2+} contributes directly to the membrane potential as a charge carrier, but also as a second messenger that controls a variety of cellular processes. These include synaptic plasticity, release of neurotransmitter, membrane exictability, enzyme activation and activity dependent gene activation (Berridge *et al.*, 2003). To selectively fulfill this multitude of important tasks, the cytosolic free calcium concentration needs to be tightly regulated in amplitude, location and duration (Augustine *et al.*, 2003).

The single compartment model is a simplified model of intracellular calcium dynamics to describe the cellular parameters that shape an intracellular Ca^{2+} signal (these include kinetics of Ca^{2+} sources, properties of intracellular Ca^{2+} buffering proteins, rate of Ca^{2+} clearance; Fig. 1.4, A). It can also be used to better understand how

1 Introduction

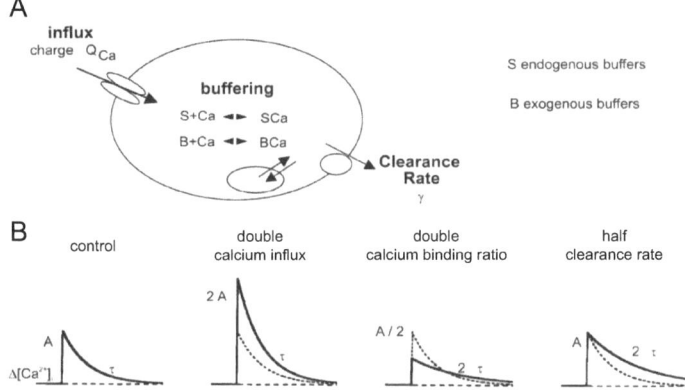

Figure 1.4: Schematic overview of the single compartment model. **(A)** Cellular parameters that shape the intracellular Ca^{2+} signal. They are mainly determined by the calcium source (influx or release from intracellular stores), the properties of intracellular Ca^{2+} buffers (calcium binding proteins) and the Ca^{2+} extrusion from the cytosol (to extracellular space and internal stores). **(B)** Simulation demonstrating how the cellular calcium handling parameters influence the actual calcium signal. A stimulus under control conditions transiently increases the intracellular Ca^{2+} concentration ($[Ca^{2+}]_i$) with a given amplitude (A) that decays to baseline with a given time constant (τ). Doubling the calcium influx results in a doubling of the signal amplitude. Doubling the Ca^{2+} binding ratio results in a bisection of the amplitude and a doubling of the time constant. Halving the clearance rate doubles the time constant of the decay. The control trace is drawn as a dashed curve for comparison. Modified from Helmchen and Tank (2005).

the relationship between influx, buffering and clearance influences the actual calcium signal (Fig. 1.4, B). This needs to be considered, because the fluorescent Ca^{2+} indicator used to measure the Ca^{2+} signal in whole-cell patch-clamp experiments acts as an added exogenous Ca^{2+} buffer that distorts the endogenous calcium buffering of the cell. To determine the endogenous Ca^{2+} handling parameters, such as endogenous decay time constant of the calcium signal, endogenous calcium binding ratio and extrusion rate, we used the 'added buffer' approach. The 'added buffer' approach is based on the single compartment model and was used in combination with patch-clamp recording and fast optical imaging (Neher and Augustine, 1992; Helmchen *et al.*, 1997; Pippow *et al.*, 2009).

1.1.5 Voltage-gated calcium channels

The main source for intracellular calcium is influx via voltage-gated calcium channels (VGCC). VGCCs are transmembrane proteins consisting of a pore forming main unit (α_1 or Ca$_v$) and auxiliary/regulatory subunits (the $\alpha_2\delta$ dimer, β and γ monomers). The main unit consists of four homologous domains, each containing six membrane segments (S1 – S6). These segments are involved in forming the pore (S5 – S6) and voltage sensing (S4; reviewed in Catterall, 2000). Many types of VGCCs have been found in vertebrates that vary in their voltage sensitivity, conductance, kinetics and pharmacology. They can be categorized broadly into two main groups by their threshold of activation: high-voltage-activated (HVA) channels that are activated at more positve membrane potentials and low-voltage-activated (LVA) channels that are activated at lower membrane potentials. A recent nomenclature proposes a classification into three structurally related families (Ca$_v$1, HVA; Ca$_v$2, HVA; Ca$_v$3, LVA; Ertel *et al.*, 2000). The Ca$_v$1-family (Ca$_v$1.1 – 1.4) comprises the historically dubbed L-type channels that are ubiquitously expressed and can be selectively blocked by 1,4-dihydropyridines, phenylalkylamines and benzothiazepines (Triggle, 2006). They play an important role in activity dependent gene regulation, contraction in muscle cells and hormone secretion. The Ca$_v$2-family (Ca$_v$2.1 – 2.3) comprises the formerly named P/Q- (Ca$_v$2.1), N- (Ca$_v$2.2) and R-type channel (Ca$_v$2.3). These channels are prominently found in neurons, especially in dendrites and nerve terminals, but also in the soma (Ca$_v$2.3). They are mostly responsible for neurotransmitter release, dendritic calcium transients and the generation of Ca^{2+}-dependent action potentials (Ca$_v$2.3). Except for R-type channels (Ca$_v$2.3), the channel types of this group can be selectively blocked by toxins from either the grass spider *Agelenopsis aperta* (ω-agatoxins; Ca$_v$2.1) or cone snails (*Conus spec.*; ω-conotoxins; Ca$_v$2.2; Randall and Tsien, 1995).

1 Introduction

The Ca_V3-family ($Ca_V3.1$ – 3.3) comprises the formerly named T-type channels. T-type channels are found in muscle cells where they play an important part in smooth muscle contraction and in neurons where they are involved in pacemaking, repetitive firing and the generation of bursts or low-threshold spikes (Perez-Reyes, 2003). There are no highly selective blockers available for T-type channels. They are, however, sensitive to organic compounds, such as amiloride and mibefradil (Tang *et al.*, 1988; Mehrke *et al.*, 1994).

1.1.6 Calcium hypothesis of brain aging

Impaired Ca^{2+} handling in neurons of the central nervous system can have dramatic consequences. They range from altered neuronal output up to neuronal death, altered behaviour or even disease (Marambaud *et al.*, 2009). Even if the changes in Ca^{2+} homeostasis are only subtle, the effect on the organism can be profound over time. This is the core idea of the 'calcium hypothesis' of neuronal aging (Khachaturian, 1987; Landfield, 1987). It proposes that during aging, the dysregulation of Ca^{2+} handling increases the Ca^{2+} load on the cell resulting in a sustained elevation of intracellular free Ca^{2+} and subsequently leading to neuronal cell loss through the neurotoxicity of Ca^{2+}. The aspect of the neurotoxicity of increased free Ca^{2+} proved to be incorrect for cognitive decline during aging but may still be valid for neurodegenerative diseases associated with aging, such as Alzheimer's (Mattson and Chan, 2003; Verkhratsky and Toescu, 2003). Nevertheless, subtle changes in Ca^{2+} homeostasis over longer periods of time are still viewed as an important factor for impaired neuronal function, synaptic activity and synaptic plasticity in old age (Toescu and Verkhratsky, 2007). Studies analyzing the Ca^{2+} homeostasis in aged animals found that in most aged neurons not only the resting Ca^{2+} level does increase, but also the Ca^{2+} influx via voltage-gated Ca^{2+} channels (Kirischuk *et al.*, 1992; Kirischuk and Verkhratsky, 1996; Murchison and Griffith, 1995; Thibault and Landfield, 1996). Further studies indicate that the Ca^{2+} buffering capacity of neurons also seems to play an important role in the cognitive decline during aging (Tonkikh *et al.*, 2006). However, the trend can vary between peripheral and central neurons or even within a region (Pottorf *et al.*, 2002; Murchison and Griffith, 1998; Murchison and Griffith, 2007; Bu *et al.*, 2003). Interestingly, caloric restriction during aging had an impact on calcium buffering properties, dampening the effects of aging on central nervous neurons (Hemond and Jaffe, 2005; Murchison and Griffith, 2007). This demonstrates a possible connection between diet and neuronal calcium homeostasis. Most studies concerning changes in calcium homeostasis and impairment of neuronal activity focus on neurons involved in memory formation and cognition. The disregulation of calcium homeostasis in neurons of the hypothalamus (e.g. POMC neurons), however, is not well understood. Moreover, it is tempting to ask whether a high caloric diet (e.g. high-fat diet) has an impact on the calcium homeostasis in POMC neurons. This, in turn, would have a dramatic impact on the organism considering the important role POMC neurons play in energy balance behaviour.

2 Methods

2.1 Animal care

Care of all animals was within institutional animal care committee guidelines. All animal procedures were approved by local government authorities (Bezirksregierung Köln, Cologne, Germany) and were in accordance with NIH guidelines. Mice were housed in groups of 3 – 5 at a temperature of 22 – 24 °C with a 12 h light/12 h dark cycle. After weaning (P21), mice were either fed regular chow food (ND; Teklad Global Rodent 2018; Harlan) containing 53.5 % carbohydrates, 18.5 % protein, and 5.5 % fat (12 % of calories from fat) or a high-fat diet (HFD; C1057; Altromin) containing 32.7 % carbohydrates, 20 % protein, and 35.5 % fat (55.2 % of calories from fat). Animals had *ad libitum* access to water and chow at all times. Calcium handling was analyzed in brain slices from P14 – P21 (3w), P105 – P140 (ND and HFD) and P304 – P334 (45w) old C57BL/6J mice expressing EGFP selectively in POMC neurons (Cowley *et al.*, 2001). The cellular properties of SIM1 neurons were determined in brain slices from C57BL/6J mice expressing EGFP selectively in SIM1 neurons (Balthasar *et al.*, 2005). All mice used in this thesis were kindly provided by the lab of Jens Brüning.

2.2 Preparation of brain slices

The animals were anesthetized with halothane (B4388; Sigma-Aldrich, Taufkirchen, Germany) and decapitated. The brain was rapidly removed from the skull cavity, cut at the brainstem and glued to a holding plate with superglue (Pattex Blitz Kleber gel, Henkel, Düsseldorf, Germany). Coronal slices (250 – 300 µm) were cut with a vibrating microtome (Microm HM 650 V; Thermo Scientific, Walldorf, Germany) in 4 °C cold, carbogenated (95 % O_2/5 % CO_2), glycerol-based, modified artificial cerebrospinal fluid to enhance the viability of neurons (GaCSF; Ye et al., 2006). GaCSF contained (in mM): 250 Glycerol, 2.5 KCl, 2 $MgCl_2$, 2 $CaCl_2$, 1.2 NaH_2PO_4, 10 4-(2-hydroxyethyl)-1-piperazineethane sulfonic acid (HEPES), 21 $NaHCO_3$, 5 Glucose, adjusted with NaOH to pH 7.2, resulting in an osmolarity of 310 mOsm. All chemicals, unless stated otherwise, were obtained from Sigma-Aldrich or Applichem (Darmstadt, Germany) in analytical grade purity. The brain slices were then transferred into carbogenated artificial cerebrospinal fluid (aCSF). First, they were kept for at least 20 min in a recovery bath at 35 °C and then kept at room temperature (24 °C, RT) for at least 30 min prior to recording. The aCSF solution contained (in mM): 125 NaCl, 2.5 KCl, 2 $MgCl_2$, 2 $CaCl_2$, 1.2 NaH_2PO_4, 10 HEPES, 21 $NaHCO_3$, 5 Glucose, adjusted with NaOH to pH 7.2, resulting in an osmolarity of 310 mOsm. During the recordings, slices were continuously superfused with aCSF at 2 – 3 ml/min. The slice was held in place with a stainless steel slice hold-down with 1 mm thread spacing (SHD-26H/10, Warner Instruments, Hamden, CT). Neurons were visualized by infrared differential interference contrast video-enhanced microscopy, with a charge-coupled device (CCD) video camera (VX55, TILL Photonics, Planegg, Germany), mounted on a fixed stage upright Olympus BX51WI microscope (Olympus, Hamburg, Germany) with 40x

LUMplan FI/IR 0.8 and 60x LUMplan FI/IR 0.9 water immersion objectives (Dodt and Zieglgänsberger, 1990). Healthy neurons were selected by the following criteria: three-dimensional, non-transparent appearance, smooth and bright membrane. EGFP expressing cells were visualized with widefield fluorescence microscopy using an X-Cite 120 illumination system (EXFO Photonic Solutions, Ontario, Canada) in combination with a Chroma 41001 fluorescence filter set (EX: HQ480/40x, BS: Q505LP, EM: HQ535/50m, Chroma, Rockingham, VT).

2.3 Electrophysiology

Membrane potential and ionic currents of POMC neurons were measured in current and voltage clamp respectively using the patch-clamp technique in the whole-cell configuration (Hamill *et al.*, 1981). Electrodes with a tip resistance between 3 – 4 MΩ were produced with a temperature controlled pipette puller (PP-830; Narishige, London, UK) from borosilicate glass capillary tubing (GB150-8P, 0.86 x 1.5 x 80 mm, Science Products, Hofheim, Germany). Whole-cell patch-clamp recordings were performed with an EPC9 patch-clamp amplifier (HEKA) controlled by the software Pulse v8.63 (HEKA). The sample interval was 10 kHz; tail currents were sampled at 20 kHz. The signal was filtered with a series combination of two short-pass Bessel filters with a cut-off frequency of 2.9 kHz and 10 kHz. Pipette and membrane capacitance were compensated using the automatic compensation circuit of the EPC9. Voltage errors due to series resistance (R_S) were minimized using the R_S compensation of the EPC9. R_S compensation was set to 60 – 80% with a time constant (τ) of 100 µs. To remove uncompensated leakage and capacitive currents, a p/6 protocol was used (Armstrong and Bezanilla, 1974). Stimulus protocols used for each set of experiments are provided in the Results. The liquid junction potential against the extracellular solution (see Neher, 1992) was calculated for each pipette solution using Igor Pro v6.01 (WaveMetrics, Inc., Lake Oswego, OR) and Patcher's Powertools plug-in for Igor Pro, v2.04 (Mendez and Würriehausen, Max-Planck-Institut für biophysikalische Chemie, Abt. Membranbiophysik, Göttingen, Germany; http://www.mpibpc.mpg.de/groups/neher/core.php?page=software) and was compensated.

Current clamp electrophysiology

For pure current clamp electrophysiology experiments, the patch pipette was filled with a solution containing (in mM): 135 K-gluconate, 10 KCl, 2 MgCl$_2$, 10 HEPES, 0.1 ethylene glycol tetraacetic acid(EGTA), 3 KATP, 0.3 NaGTP adjusted to pH 7.2 with KOH resulting in $\sim$ 300 mOsm. Tolbutamide (200 µm; Sigma) was dissolved in dimethyl sulfoxide (DMSO) and added to the normal aCSF with a final DMSO concentration of 0.25 %. The DMSO concentration had no obvious effect on the investigated neurons.

Calcium currents

To isolate Ca^{2+} currents, a combination of pharmacological blockers and ion substitution was used. The extracellular solution for measuring Ca^{2+} currents contained (in mM): 103 choline-Cl, 2.5 KCl, 2 MgCl$_2$, 3 CaCl$_2$, 1.2 NaH$_2$PO$_4$, 10 HEPES, 21 NaHCO$_3$, 5 Glucose, 0.001 tetrodotoxin (TTX, T-550, Alomone, Jerusalem, Israel), 4 4-aminopyridine (4-AP), 20 tetraethylammonium (TEA)-Cl, 0.1 Picrotoxin (PTX), 0.05 D-2-amino-5-phosphonopentanoate (D-AP5), 0.01 6-cyano-7-nitroquinoxaline-2,3-dione (CNQX) adjusted with NaOH to pH 7.2, resulting in an osmolarity of 310 mOsm. The pipette solution contained in (mM): 133 CsCl, 2 MgCl$_2$, 1 CaCl$_2$, 10 HEPES, 10 EGTA, 3 KATP, 0.3 NaGTP adjusted to pH 7.2 with KOH, resulting in $\sim$ 290 mOsm.

Calcium imaging

For Ca^{2+} imaging experiments, the extracellular solution contained (in mM): 125 NaCl, 2.5 KCl, 2 MgCl$_2$, 2 CaCl$_2$, 1.2 NaH$_2$PO$_4$, 10 HEPES, 21 NaHCO$_3$, 5 Glucose, 0.1 PTX, 0.05 D-AP5, 0.01 CNQX adjusted with NaOH to pH 7.2, resulting in an osmolarity of 310 mOsm. The pipette solution contained (in mM): 135 K-gluconate, 10 KCl, 2 MgCl$_2$, 10 HEPES, 0.2 fura-2 (F1200, Invitrogen, OR, USA) adjusted to pH 7.2 with KOH, resulting in $\sim$ 300 mOsm.

Data analysis

Steady-state, tail current activation data of the calcium currents were fit using a first-order and second-order (n = 1,2) Boltzmann equation:

$$\frac{I}{I_{max}} = \frac{1}{(1+e^{\frac{V-V_{0.5}}{s}})^n} \quad (2.1)$$

Steady-state inactivation data were fit using a sum of two first-order Boltzmann equations (Wicher and Penzlin, 1997; Heidel and Pflüger, 2006):

$$\frac{I_i}{I_{i,max}} = \frac{1}{(1+e^{\frac{V-V_{i,0.5}}{s_i}})} \quad , i = 1,2 \quad (2.2)$$

I_{max} is the maximal current, V is the voltage of the test pulse, I is the current at voltage V, s is the slope factor and $V_{0.5}$ is the voltage at which half maximal activation occurs. To describe the time course of calcium current inactivation kinetics, a sum of exponential functions with up to two time constants (i = 1,2) was used:

$$I_i = A_i \cdot e^{\frac{-t}{\tau_i}} \quad (2.3)$$

Where τ_i is the time constant, A_i is the current amplitude at $t = 0$ and t is the time at which the current I_i occurs. The recorded data was analyzed using the software Pulse, Igor Pro v6.01 (WaveMetrics), Patcher's Powertools plug-in for Igor Pro and Sigma Stat (version 3.1; Systat Software Inc.). To determine differences in the mean values between the different cohorts, ANOVA and post hoc Tukey's multiple comparison tests were performed. Significance was accepted at $p \leq 0.05$. All calculated values are expressed as mean ± standard deviation (SD). Box plot creation and statistical analysis were performed with GraphPad Prism version 5.01 (GraphPad Software, La Jolla, CA).

Current clamp data were analyzed using the NeuroMatic plug-in for Igor Pro (Jason Rothman, http://www.neuromatic.thinkrandom.com/index.html) and Igor Pro. Significance of differences between mean values was evaluated with paired and unpaired t-tests. A significance level of 0.05 was accepted for all tests. To determine if there was an effect of tolbutamide on the membrane potential the 3SD criterion was used. A neuron was considered tolbutamide sensitive when the depolarization during tolbutamide application was > 3SD of the control membrane potential and the effect was reversible during wash-out (Kloppenburg et al., 2007; Ernst et al., 2009).

2.4 Fluorimetric calcium measurements

2.4.1 Imaging setup

The setup consisted of a fixed stage upright microscope (Olympus BX51WI) with a 20x XLUMPlan Fl/0.95 NA water immersion objective and a C-Mount TV Magnifier (2x and 4x magnification for the camera). For fluorescence excitation a Polychrome IV Monochromator Xenon lamp (TILL-Photonics GmbH, Gräfeling, Germany) was used.

Intracellular Ca^{2+} concentrations were measured with the Ca^{2+} indicator fura-2. The neurons were loaded with fura-2 via the patch pipette and illuminated during data collection with light from the polychromator (with a wavelength of 340 nm, 360 nm or 380 nm), reflected onto the cell by a 410 nm dichroic mirror (DCLP410, Chroma, Gräfeling, Germany). Emitted fluorescence was detected through a 440 nm long-pass filter (LP440, Chroma) by an IMAGO 12-bit VGA CCD-camera (TILL-Photonics). The camera and monochromator were controlled via TILLvisION software v4.0 (TILL-Photonics). Data were acquired as 40x30 pixel frames using 8x8 on-chip binning. Images were recorded in analog-to-digital units (ADUs) and stored and analyzed as 12-bit grayscale images. To calculate the kinetics, the mean value of ADU within regions of interest (ROIs) from the center of the soma were used. ROIs were adjusted to each cell. For background subtraction, a second ROI containing unmarked tissue was chosen. To correct for increased fluorescence levels of the ROI positioned over the cell body compared to the ROI from the background (possibly due to EGFP fluorescence), a background correction factor was calculated for each excitation wavelength by dividing the mean ADU of the cellbody before break-in over the mean ADU of the background before break-in. The background fluorescence was multiplied by the correction factor and was then subtracted from the fura-2 fluorescence (see Lee *et al.*, 2000, Fig. 2.1).

2.4.2 Experimental procedures

After establishing the whole-cell configuration, neurons were voltage clamped at -80 mV and intracellular dye loading was monitored at 360 nm excitation, the isosbestic point of fura-2. Frames were taken at 30 s intervals (7 ms exposure time). The intracellular fura-2 concentration was estimated for different times during the loading curve, assuming that cells were fully loaded when the fluorescence reached a plateau and that the fura-2 concentration in the cell and in the pipette is the same (200 µM; Lee et al., 2000). To measure the resting calcium level of the neuron immediately after break-in and to monitor the physiological status of the cell, loading curves were also measured at 340 and 380 nm excitation. During fura-2 loading, voltage-activated Ca^{2+} influx was induced by stepping the voltage-clamped membrane potential to 10 mV for 100 ms. To monitor the elicited Ca^{2+} transients ratiometrically, pairs of images at 340 nm (15 ms exposure time) and 380 nm (6 ms exposure time) excitation were acquired at 13.3 Hz for 12 s. Typically, three Ca^{2+} transients were elicited during loading at different intracellular fura-2 concentrations approximately 300, 600 and 1200 s after establishing the whole-cell configuration (see also Results, Fig. 3.9).

2.4.3 Data analysis of fluorimetric calcium measurements

Calibration of fura-2

To determine the intrinsic buffering and extrusion values for POMC neurons, the 'added buffer' approach was used (Neher and Augustine, 1992; Helmchen et al., 1997). Procedures and data analysis were adapted from Pippow et al. (2009) and are briefly described here. For a detailed description, please refer to the original publication. Calibration constants for fura-2 were determined according to Grynkiewicz et al. (1985),

2 Methods

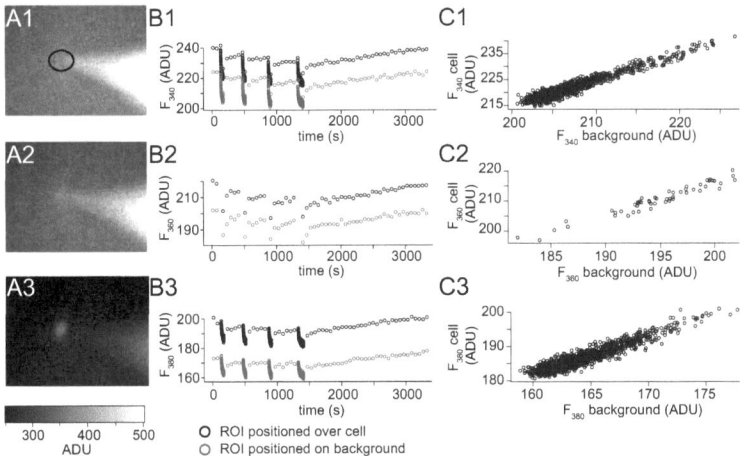

Figure 2.1: Background fluorescence subtraction. (A1 – 3) Fluorescence images of the recording situation with 340 nm (A1), 360 nm (A2) and 380 nm (A3) excitation wavelength. The black circle in A1 displays the region of interest over the soma of the neuron, the grey circle displays the ROI over unmarked tissue. Note that EGFP fluorescence could be detected with 380 nm excitation wavelength (A3). **(B1 – 3)** Mean fluorescence intensity over both ROIs with 340 nm (A1), 360 nm (A2) and 380 nm (A3) excitation wavelength plotted over time. Black trace representing the ROI over the soma, grey trace the ROI over the unmarked tissue. The trace was recorded during a 'blank' experiment where the same experimental procedure for the Ca^{2+} imaging experiments was followed but without fura-2 in the pipette solution. Note that there is a linear relationship between the background fluorescence and the fluorescence of the cell. **(C)** Plot of the mean fluorescence of the ROI over the cell against the mean fluorescence of the ROI over the unmarked tissue with 340 nm (A1), 360 nm (A2) and 380 nm (A3) excitation wavelength. The linear relationship allows to determine a correction factor by dividing the mean ADU of the cellbody before break-in over the mean ADU of the background before break-in.

2 Methods

Poenie (1990) and Zhou and Neher (1993) as described in Pippow *et al.* (2009).

The ability of EGTA to bind calcium is highly dependent on the environmental conditions such as ionic strength, temperature, pH and the concentrations of other metals that compete for binding (Harrison and Bers, 1987; Harrison and Bers, 1989). In theory the necessary amount of Ca^{2+} and EGTA to set the free Ca^{2+} concentration for the experimental conditions can be calculated (Patton *et al.*, 2004). However, small variations in the parameters such as pH, temperature, impurities of chemicals, pipetting or weighing errors can lead to considerable errors when estimating the free Ca^{2+} concentration in EGTA-buffered Ca^{2+} solutions with computer programs (McGuigan *et al.*, 2007). To account for such variations, the free Ca^{2+} concentration in the calibration solutions was determined by using a Ca^{2+} selective electrode following the guide from McGuigan *et al.* (1991). Calibration solutions were prepared as follows (in mM): Rmax: 140 KCl, 2.5 KOH, 15 NaCl, 1 $MgCl_2$, 5 HEPES, 10 $CaCl_2$ and 0.05 fura-2; Rmin: 129.5 KCl, 13 KOH, 15 NaCl, 1 $MgCl_2$, 5 HEPES, 4 EGTA and 0.05 fura-2; Rdef: 129.5 KCl, 13 KOH, 10.3 NaCl, 4.7 NaOH, 1 $MgCl_2$, 5 HEPES, 4 EGTA, 2.7 $CaCl_2$ and 0.05 fura-2, yielding a free Ca^{2+} concentration of 0.35 µM. All solutions were adjusted to pH 7.2 with HCl.

The measured fluorescence ratio R can be converted to the intracellular calcium concentration ($[Ca^{2+}]_i$) using:

$$[Ca^{2+}]_i \;=\; K_{d,fura,eff}\, \frac{R - R_{min}P}{R_{max}P - R} \tag{2.4}$$

Where R_{max} is the fluorescence ratio at saturating Ca^{2+} concentrations and R_{min} the ratio under Ca^{2+} free conditions. P is a correction factor for R_{max} and R_{min} to account for differences between cytosol and the calibration solution (see Poenie, 1990). $K_{d,fura,eff}$ is the effective dissociation constant of fura-2, independent of the dye concentration and specific for each experimental setup (see Neher, 1989). $K_{d,fura,eff}$ is defined as:

$$K_{d,fura,eff} = [Ca^{2+}]_i \frac{R_{max} - R}{R - R_{min}} \quad (2.5)$$

To determine the endogenous Ca^{2+} binding ratio, the dissociation constant ($K_{d,fura}$) was by calculated using (Zhou and Neher, 1993):

$$K_{d,fura} = K_{d,fura,eff} \frac{R_{min} + \alpha}{R_{max} + \alpha} \quad (2.6)$$

To compensate differences in signal intensity at 340 and 380 nm excitation of fura-2, the isocoefficient factor α is also needed to calculate $K_{d,fura}$. It is the factor for which the sum:

$$F_i(t) = F_{340}(t) + \alpha F_{380}(t) \quad (2.7)$$

is independent of the Ca^{2+} concentration (Fig. 2.2). F_{340} and F_{380} are background subtracted fluorescence signals measured during a brief Ca^{2+} transient.

In the cell Ca^{2+} is bound to the Ca^{2+} buffers fura-2 and to the immobile endogenous buffer, which are both assumed to be always in equilibrium with free Ca^{2+} and unsaturated. The ability of the experimentally introduced exogenous buffer to bind Ca^{2+} is described by its Ca^{2+} binding ratio that is defined as the ratio of the change in buffer-bound Ca^{2+} over the change in free Ca^{2+}:

$$\kappa_B = \frac{d[BCa]}{d[Ca^{2+}]_i} = \frac{[B_T]K_{d,B}}{([Ca^{2+}]_i + K_{d,B})^2} \quad (2.8)$$

Where $[BCa]$ is the concentration of the exogenous buffer (fura-2) in its Ca^{2+} bound form, $[B_T]$ is the total concentration of the exogenous buffer B and $K_{d,B}$ is its dissociation constant for Ca^{2+}.

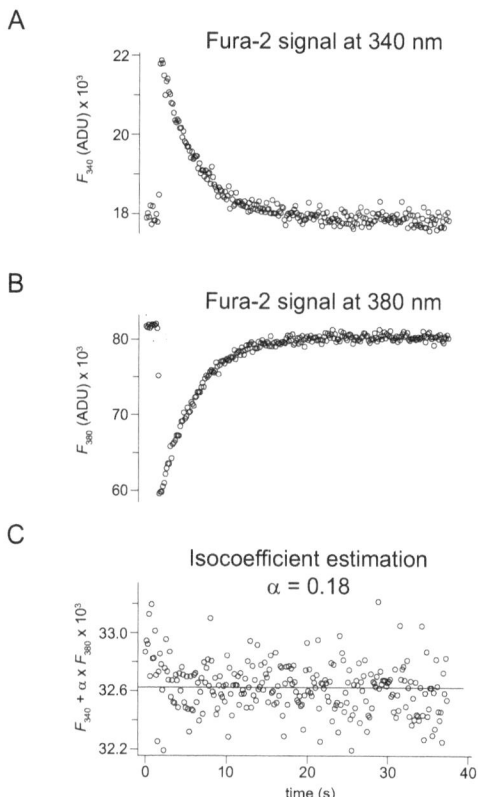

Figure 2.2: Estimation of the isocoefficient α. The isocoefficient α must be known to calculate the dissociation constant of the Ca^{2+} indicator fura-2 ($K_{d,fura}$; Equation 2.6). The isocoefficient is the factor for which $F_i(t) = F_{340}(t) + \alpha F_{380}(t)$ is independent of $[Ca]_i$. F_{340} and F_{380} are the fluorescence signals for a Ca^{2+} transient measured **(A)** at 340 and **(B)** at 380 nm excitation, respectively. **(C)** α is determined by iteration, optimizing $F_i(t)$ to a horizontal line for the decay of the signals. t = 0 s is the start of the fluorescence signal.

2 Methods

Single compartment model of calcium buffering

For measurements of intracellular Ca^{2+} concentrations with Ca^{2+} chelator-based indicators, the amplitude and time course of the signals are dependent on the concentration of the Ca^{2+} indicator (in this case fura-2). Fura-2 enters the cell via the patch pipette and acts as an exogenous (added) Ca^{2+} buffer, competing with the endogenous buffer (Neher and Augustine, 1992; Tank et al., 1995; Helmchen et al., 1997; Helmchen and Tank, 2005). With the 'added buffer' approach (Neher and Augustine, 1992), the capacity of the immobile endogenous Ca^{2+} buffer in a cell is determined by measuring the decay of the Ca^{2+} signal at different concentrations of 'added buffer' and by extrapolating to conditions in which only the endogenous buffer is present.

Assuming a constant $[Ca^{2+}]_i$, a constant exogenous buffer concentration and Ca^{2+} transients smaller than $0.5\,K_{d,fura}$ (see Neher and Augustine, 1992; Tank et al., 1995; Pippow et al., 2009) the model describes the decay of a voltage induced Ca^{2+} transient as:

$$\tau_{transient} = \frac{1 + \kappa_B + \kappa_S}{\gamma - ((1 + \kappa_S)/\tau_{loading})} \quad (2.9)$$

$\tau_{transient}$ is proportional to κ_B. A linear fit to the data has its negative x-axis intercept at $1 + \kappa_B$, yielding the endogenous Ca^{2+} binding ratio of the cell. The slope of this fit is the inverse Ca^{2+} extrusion rate γ. The y-axis intercept yields the time constant τ_{endo} for the decay of the Ca^{2+} transient as it would appear in the cell without exogenous buffer.

Fitting the loading curve

Fluorescence was measured at the isosbestic wavelength of fura-2 to monitor loading of fura-2 into the neurons. The loading curves were fit with the exponential function:

$$Y = Y_{max}(1 - e^{(-(t-t_{bi})/\tau_{loading})}) \qquad (2.10)$$

Y is the fluorescence intensity in ADU or, after rescaling, the concentration of fura-2. Y_{max} is the plateau of fluorescence or maximum concentration of fura-2 (200 µM), $\tau_{loading}$ is the loading time constant, t_{bi} is the time of break-in. Initially, the data were fit with Eq. 2.10 using the least squares method implemented by the 'R function' nls (Nonlinear Least Squares, R Development Core Team; 2007, R: A language environment for statistical computing; R Foundation for Statistical Computing, Vienna, Austria; http://www.R-project.org). Reasonable 'initial guesses' were obtained by the 'R function' SSasympOff (Asymptotic Regression Model with an Offset, R Development Core Team; 2007). To determine the absolute fura-2 concentration during the time course of the experiment the fluorescence intensity was scaled to fura-2 concentrations (Y_{max} = 200 µM) and the rescaled data were fit again with Eq. 2.10. This fitting procedure describes the overall loading process very well. On a shorter time scale, however, the data sometimes fluctuated around the fit (Fig. 2.3, A), probably due to changes in access resistance (Helmchen *et al.*, 1996). Improved estimates of the fura-2 concentration at any time throughout the loading could be obtained from subsequent smoothing spline fits (Fig. 2.3, B; Green and Silverman, 2000), which were computed with the 'R function' smooth.spline (R Development Core Team; 2007). For the analysis of calcium buffering, the fura-2 concentrations obtained from the smoothing spline fit were used.

2 Methods

Fitting the transients

Fluorescence ratios (R) from 340 and 380 nm excitation after a brief voltage-activated Ca^{2+} influx were recorded to monitor the time course of $[Ca^{2+}]_i$ transients. Values for each wavelength were acquired in ADU from a ROI from the center of the soma. The ADU counts of each wavelength were used to compute ratios that were then transformed to absolute Ca^{2+} concentrations as described above. The decay of the transient was fit with the monoexponential function:

$$[Ca^{2+}]_i(t) = S_{drift}t + [Ca^{2+}]_{i,0}\ e^{-t/\tau_{transient}} \tag{2.11}$$

$S_{drift}t$ is a linear drift term taking bleaching into account (with S_{drift} for the slope and t for time), $[Ca^{2+}]_{i,0}$ the amplitude of the signal at its peak and τ is the decay constant. Parameters of Eq. 2.11 were estimated using the 'R function' nls (R Development Core Team; 2007). Starting values for S_{drift} were obtained from the slope of the baseline, for $[Ca^{2+}]_{i,0}$ by subtracting the baseline of the signal from the peak amplitude, and for τ from the time point where the amplitude had decreased e-fold.

Fitting the linearized model

Time constants of transients were plotted as a function of κ_B values and fit with a linear function:

$$Y = \beta_0 + \beta_1 x \tag{2.12}$$

using the 'R function' lm (R Development Core Team; 2007). To estimate the variance of the slope (extrusion rate γ), the y-axis intercept (endogenous decay constant τ) and the negative x-axis intercept (endogenous Ca^{2+} binding ratio κ_S), we used the bootstrap method (Efron, 1979; Aponte et al., 2008) implemented in the boot library in R (fixed-x resampling, 1000 bootstrap samples, boot: Bootstrap R Functions, R package version 1.2-27). This resulted in bootstrap distributions (n = 1000) for each of the parameters (i.e. extrusion rate γ, endogenous decay constant τ and endogenous Ca^{2+} binding ratio κ_S). The distributions were *log*-transformed to make them closer to a Gaussian. To determine differences in the mean values between the different cohorts, ANOVA and post hoc pairwise comparisons were performed using t-tests with the Holm method for p-value adjustment. Significance was accepted at $P \leq 0.05$. All calculated values are expressed as mean $\pm$ SD.

Calibration of the Ca^{2+} selective electrode

To calibrate the Ca^{2+} selective electrode, the approach described by McGuigan *et al.* (1991) was followed. Five electrode calibration solutions with increasing Ca^{2+} concentration were prepared containing (in mM): 140 KCl, 2.5 KOH, 15 NaCl, 1 $MgCl_2$, 5 HEPES and 10, 4, 1, 0.4, 0.2 $CaCl_2$, respectively. Additionally, the following solutions were prepared: an 'EGTA solution' containing (in mM): 129.5 KCl, 13 KOH, 15 NaCl, 1 $MgCl_2$, 5 HEPES, 4 EGTA and a 'Ca-EGTA solution' containing (in mM): 129.5 KCl, 13 KOH, 8 NaCl, 7 NaOH, 1 $MgCl_2$, 5 HEPES, 4 EGTA, 4 $CaCl_2$. All solutions were adjusted to pH 7.2 with HCl. Twelve EGTA/Ca-EGTA solutions were mixed in ratios of: 1:10, 1:9, 1:8, 1:7, 1:6, 1:5, 1:4, 1:3, 1:2, 3:1, 5:1, 9:1. Using a Ca^{2+} selective macroelectrode (Ca 800, Wissenschaftlich- Technische Werkstätten GmbH, Weilheim, Germany), the potentials of the electrode calibration solutions, as well as the EGTA/Ca-EGTA mixtures, were measured in order from the highest to the lowest free Ca^{2+} concentration. After measuring the last solution, the potential of the initial solution was measured again to check for a possible drift of the macroelectrode in which case the potentials were drift corrected. The measured potentials in the Ca^{2+} calibration solutions can be

2 Methods

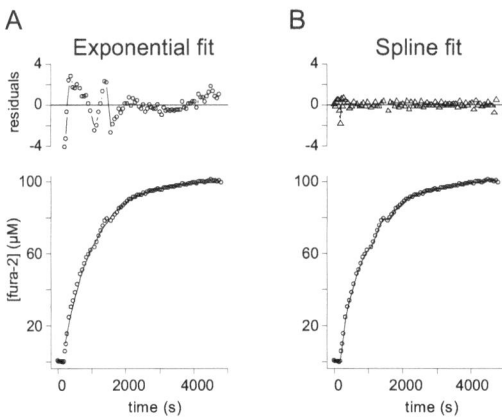

Figure 2.3: Fitting the loading curves. Loading curves were acquired as in Fig. 3.9. The data (in ADU) were fit with a monoexponential function (Eq. 2.10). To obtain the fura-2 concentration during measurement of the transients, the data were scaled to fura-2 concentrations with $Y_{max} = 200\,\mu M$, the fura-2 concentration in the recording pipette. The scaled data were fit again with Eq. 2.10. This fit (shown in **A**) describes the overall loading process very well. On a shorter timescale, however, the data sometimes fluctuated around the fit, propably caused by a changing R_s. Improved, time resolved estimates of the fura-2 concentrations could be obtained from a subsequent smoothing spline fit (shown in **B**). The residuals demonstrate the improvement of the fit.

2 Methods

described by the Nernst equation (Nernst, 1888) and can thus be used to determine the slope of the electrode (change in potential per change in 10-fold Ca^{2+} concentration in mV · pCa^{-1}). In contrast, the potentials of the buffer solutions cannot be described by the Nernst equation, because their free Ca^{2+} concentrations are in the nanomolar range, where the slope of the electrode becomes nonlinear. In this case the Nikolsky-Eisenmann equation relates the measured potentials in the buffer solutions to free Ca^{2+} concentrations (Kay et al., 2008):

$$E = E^0 + s\,log\left(\left[X^{2+}\right] + \Sigma\right) \tag{2.13}$$

E is the measured potential of the electrode in mV, E^0 is the standard electrode potential in mV, s is the slope of the electrode in mV · pCa^{-1}, $\left[X^{2+}\right]$ is the free Ca^{2+} concentration in mol · l^{-1} and Σ is an interference constant accounting for nonlinearity of the slope in the nanomolar range in mol · l^{-1}. $\left[X^{2+}\right]$ can be calculated from the total added Ca^{2+} concentration $[X]_T$, the total EGTA concentration $[Ligand]_T$ and the EGTA dissociation constant K_{app}:

$$\left[X^{2+}\right] = \frac{b + \sqrt{b^2 + 4K_{app}\,[X]_T}}{2} \tag{2.14}$$

where,

$$b = -\left(K_{app} + [Ligand]_T - [X]_T\right) \tag{2.15}$$

There are three unknown parameters, which must be calculated, namely $[Ligand]_T$, K_{app}, and Σ in the Nicolsky-Eisenman equation. In Luthi et al. (1997), a cyclic iterative scheme was used to calculate the $[Ligand]_T$ and K_{app} parameters, but a fixed value for the parameter Σ (see McGuigan et al., 2006 for details) was used. When these parameters are known, the free Ca^{2+} concentration of the buffer solutions can be calculated with the Nikolsky-Eisenman equation. All calculations were performed in R using the 'R function' ALE (Automatic determination of Ligand purity and Equilibrium dissociation constant; Kay *et al.*, 2008).

2.5 Single cell labeling and microscopy

To label single cells, 0.1 % biocytin (B4261; Sigma-Aldrich) was added to the pipette solution (Horikawa and Armstrong, 1988). After the recordings, the slices were fixed in Roti-Histofix (P0873; Carl Roth) overnight at 4 °C and rinsed in 0.1 M 2-amino-2-(hydroxymethyl)-1,3-propanediol (TRIS)-HCl-buffered solution (TBS), pH 7.2 (three times for 10 min each time). To facilitate the streptavidin penetration, the slices were incubated in TBS containing 1 % Triton X-100 (30 min; RT; Serva). Afterwards, the slices were washed in TBS (3x, RT) and incubated in Alexa Fluor 633 (Alexa 633)-conjugated streptavidin (1:600; 1 – 2 d; 4 °C; S21375; Invitrogen) that was dissolved in TBS containing 10 % normal goat serum (S-1000; Vector Labs). The slices were rinsed in TBS (three times for 10 min each and one over night at 4 °C), dehydrated in an alcohol series consisting of 50, 70, 90 and 2 times 100 % ethanol (Serva), treated with xylene for 10 min, and mounted in Permount (SP15B; Thermo Fisher Scientific).

Fluorescence images of the labeled neurons were captured with a confocal microscope (LSM 510; Carl Zeiss) equipped with a Plan-Neofluar 10x 0.3 NA objective. Streptavidin-Alexa 633 was excited at 633 nm using a Helium-Neon laser. Emission of Alexa 633 was collected through a 650 nm LP filter. For overviews, overlapping stacks were acquired and merged using the photomerge function in Photoshop CS2 (Adobe Systems). Calibration, noise reduction, and z-projections were done using ImageJ, version 1.42i, with the WCIF plugin bundle (http://www.uhnresearch.ca/facilities/

wcif/).

3 Results

The main objective of this thesis was to better understand the cellular properties of identified hypothalamic neurons that regulate energy homeostasis and in addition, set the stage for future experiments. The goal was approached in two parts.

First, to better understand the calcium handling in first-order neurons of the energy homeostatic network (POMC neurons) and how it is influenced during maturation by a high-fat diet, the calcium current in POMC neurons was characterized with whole-cell voltage clamp recordings. Furthermore, the calcium buffering capacity and extrusion rate of POMC neurons using the 'added buffer' approach were determined. This was achieved by performing ratiometric calcium imaging experiments in combination with whole-cell recordings of identified POMC neurons.

Second, potential candidates for second-order neurons to POMC neurons in the PVH were characterized, performing whole-cell patch-clamp current clamp recordings in a genetically identified subpopulation of PVH neurons: SIM1 neurons.

3.1 POMC neuron morphology

To characterize their morphological properties, POMC neurons expressing EGFP were recorded with the whole-cell patch-clamp technique in the living brain slice, filled with biocytin and stained after fixation with Alexa633-streptavidin (n=35). The morphology varied substantially between recorded POMC neurons (Fig. 3.1). In some cases the neurons had dendritic arborizations in the ARC and extended axonal projections in a medio-lateral or ventral direction (Fig. 3.1, A). In other cases, especially neurons located around the median eminence and close to the third ventricle, showed very sparse or no dendritic arborization (Fig. 3.1, B). Despite this morphological heterogeneity, apparent differences in the electrophysiological responses in current and voltage clamp recordings could not be detected using our experimental paradigms. Therefore, the Ca^{2+} currents that are presented in the following section were pooled for analysis.

3 Results

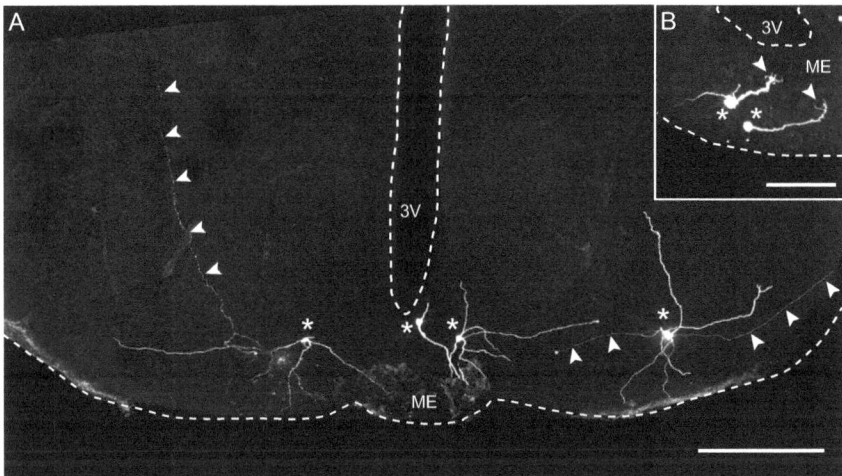

Figure 3.1: Morphology of POMC neurons. **(A)** Maximum intensity projection of a confocal image stack. Morphology of exemplary POMC neurons that were stained with biocytin/Alexa Fluor 633-streptavidin. Note that the neurons qualitatively differ in their dendritic arborization pattern and the direction of their thin axonal projecting processes (arrowheads) or lack thereof. **(B)** Morphology of two representative POMC neurons located around the median eminence. Note that the neurons qualitatively lack elaborate dendritic arborizations but display relatively short and broad processes which end in short and diffuse terminals (arrowheads). The somata of the neurons are marked by an asterisk (*). Scale bar in A: 250 µm, in B: 100 µm. 3V: third ventricle, ME: median eminence.

3.2 Voltage-activated calcium currents in POMC neurons

Voltage-activated calcium currents (I_{Ca}) from a total of 142 POMC neurons were recorded with the patch-clamp technique in the whole-cell voltage clamp configuration. Other ionic currents were reduced by application of known specific blockers and ion substitution (Hille, 2001; see Methods). Given that POMC neurons have a relatively complex morphology (Fig. 3.1), proper voltage clamp is difficult to achieve (Armstrong and Gilly, 1992; White et al., 1995). However, recordings did not indicate poor voltage clamp (delay of current onset, jumps in the voltage dependence), suggesting that the recorded currents originated from well clamped regions of the neuron. Recordings with high access resistances (> 16 MΩ) were discarded.

I_{Ca} was recorded in POMC neurons of the arcuate nucleus (ARC) from four different cohorts: P14 – 21 suckling milk (3w), P105 – 140 normal diet (ND), P105 – 140 high-fat diet (HFD) and P304 – 334 normal diet (45w) mice. For these cohorts, steady-state activation, capacitance, current density, steady-state tail current activation, steady-state inactivation and inactivation kinetics were determined and compared to detect possible age and diet dependent changes of I_{Ca} in POMC neurons. For steady-state activation, current density and steady-state tail current activation, the overall current was activated from a holding potential (E_{hold}) of -100 mV and from E_{hold} -50 mV to isolate the high-voltage-activated (HVA) currents.

3.2.1 Current/voltage relationship

To determine the I/V relationship, I_{Ca} was elicited with steady state activation and tail current activation protocols.

Steady-state activation

To measure steady state activation, the neurons were depolarized for 50 ms from E_{hold} to 50 mV in 5 mV steps. During a depolarizing pulse, the total current activated quickly and decayed relatively slowly (Fig. 3.2, A, E_{hold} -100 mV). The HVA currents also activated quickly but had a smaller amplitude and a slower decay (Fig. 3.2, A, E_{hold} -50 mV). The capacitance for each neuron was measured, allowing to calculate the current density (Fig. 3.2, B). Compared to all other cohorts, the HFD cohort showed a greater maximal peak current density of the overall current (Fig. 3.3, A1 & A2) and of the HVA currents, indicating an increase of voltage-activated Ca^{2+} influx as a result of the high-fat diet (Fig. 3.3, B1 & B2). For the total current, the command potential where activation of the peak current amplitude is maximal (V_{max}) was more depolarized in the 3w cohort compared to all other cohorts (Fig. 3.3, C). This indicates that for the total current of I_{Ca}, but not for HVA currents, there is a slight shift towards a more depolarized V_{max} during the transition from suckling to weaned mice. For HVA currents, V_{max} was more hyperpolarized in the 45w cohort compared to all other cohorts (Fig. 3.3, D). This suggests that there is a shift in V_{max} of HVA currents towards a more hyperpolarized potential during the aging from young to mature (P304 – 334) mice.

The command potentials at which the currents activated (-70 mV for the total current, -50 mV for the HVA currents) were not significantly different in all cohorts. The values of the steady-state activation parameters of I_{Ca} for all cohorts are summarized in Tab. 3.1.

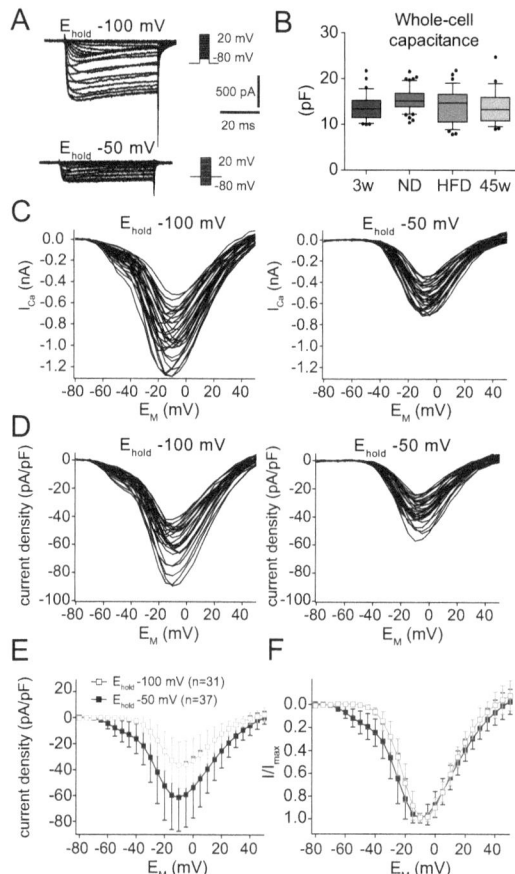

Figure 3.2: Steady-state activation of voltage-activated calcium currents (I_{Ca}) measured in POMC neurons of mice from the ND cohort. (A) Example current traces for steady-state activation of I_{Ca} from a holding potential (E_{hold}) of -100 mV and -50 mV respectively. The neurons were depolarized for 50 ms to at least 20 mV in 5 mV increments. (B) Cell capacitance was not significantly different between cohorts. For capacitance mean values see Tab. 3.1. (C) Voltage dependence of peak I_{Ca} for the ND cohort representative for the other cohorts. Absolute current data from single neurons with E_{hold} -100 mV (n=31) and -50 mV (n=37). (D) Current density/voltage relation representative for the other cohorts. Current density was calculated from the ratio of I_{Ca} and the capacitance of the neuron. Absolute current data from single neurons with E_{hold} -100 mV and -50 mV. (E) Current density/voltage relation with E_{hold} -100 mV and E_{hold} -50 mV. Averaged data. The mean maximal current density with E_{hold} -100 mV was -59.63 ± 10.74 mV and with E_{hold} -50 mV, -36.77 ± 9.08 mV. (F) I/V relation of peak I_{Ca} normalized to the peak current of each cell with E_{hold} -100 mV and E_{hold} -50 mV. For E_{hold} -100 mV, the current is activated at command potentials more depolarized than -70 mV with a maximum at -10.94 ± 3.46 mV and with E_{hold} -50 mV at command potentials more depolarized than -50 mV with a maxmium at -8.03 ± 2.48 mV. Averaged data. Error bars in E and F show SD.

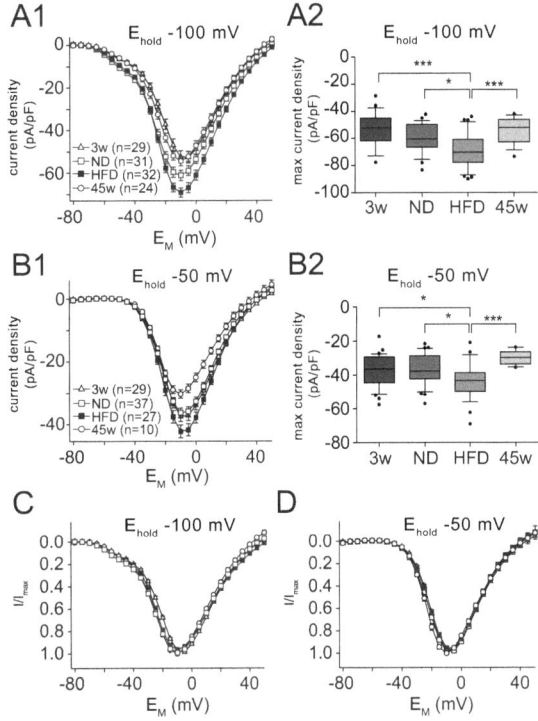

Figure 3.3: I_{Ca} current density and I/V relation in different age and diet cohorts. (**A1**) Current density/voltage relation of I_{Ca} recorded from P14 – 21 suckling milk (3w), P105 – 140 normal diet (ND), P105 – 140 high-fat diet (HFD) and P304 – 334 normal diet (45w) mice with E_{hold} -100 mV. Averaged data. In all cohorts, the current is activated at command potentials more depolarized than -70 mV. (**A2**) Mean maximal current densities with E_{hold} -100 mV were as follows: 3w, -54.24 ± 14.87 pA/pF; ND, -59.63 ± 10.74 pA/pF; HFD, -69.50 ± 12.23 pA/pF; 45w, -53.95 ± 9.48 pA/pF. The increase of the mean maximal current density of the HFD cohort compared to all other cohorts was statistically significant. (**B1**) Current density/voltage relation for the same cohorts with E_{hold} -50 mV. (**B2**) Mean maximal current densities with E_{hold} -50 mV were as follows: 3w, -36.95 ± 9.22 pA/pF; ND, -36.77 ± 9.08 pA/pF; HFD, -43.59 ± 10.40 pA/pF; 45w, -29.79 ± 3.874 pA/pF. The increase of the mean maximal current density of the HFD cohort compared to all other cohorts was statistically significant. (**C**) I/V relation of peak I_{Ca} normalized to the maximal current of each cell with E_{hold} -100 mV. Averaged data. In all cohorts, the current is activated at command potentials more depolarized than -70 mV. The values for the command voltage where the peak current reached its maximum (V_{max}) were as follows: 3w, -7.41 ± 2.87 mV; ND, -10.94 ± 3.46 mV; HFD, -10.47 ± 4.47 mV; 45w, -10.56 ± 2.92 mV. In the 3w cohort, V_{max} is significantly more positive compared to all other cohorts (3w vs ND,HFD: $p < 0.01$; 3w vs 45w: $p < 0.05$). (**D**) I/V relation of peak I_{Ca} normalized to the maximal current of each cell with E_{hold} -50 mV. Averaged data. In all cohorts, the current is activated at command potentials more depolarized than -50 mV. The values for V_{max} were as follows: 3w, -7.57 ± 2.79 mV; ND, -8.06 ± 2.47 mV; HFD, -7.78 ± 2.53 mV; 45w, -10.50 ± 1.58 mV. In the 45w cohort, V_{max} is significantly more negative compared to all other cohorts (45w vs 3w, $p < 0.01$; 45w vs ND,HFD, $p < 0.05$). Error bars in A1, B1, C and D show standard error of mean (SEM). Asterisks mark the level of significance: *** $p < 0.001$, * $p < 0.05$.

Table 3.1: Summary of steady-state activation parameters of I_{Ca}.

	3w	ND	HFD	45w
Whole-cell capacitance (pF)	13.67 ± 2.86 (n=33)	15.42 ± 2.43 (n=50)	13.94 ± 3.8 (n=35)	13.97 ± 3.78 (n=24)
Max current density E_{hold} -100mV (pA/pF)	-54.24 ± 14.87 (n=29)	-59.63 ± 10.74 (n=31)	-69.50 ± 12.23 (n=32)	-53.95 ± 9.48 (n=24)
Max current density E_{hold} -50mV (pA/pF)	-36.95 ± 9.22 (n=29)	-36.77 ± 9.08 (n=37)	-43.59 ± 10.40 (n=27)	-29.79 ± 3.87 (n=10)
V_{max} E_{hold} -100mV (mV)	-7.41 ± 2.87	-10.94 ± 3.46	-10.47 ± 4.47	-10.56 ± 2.92
V_{max} E_{hold} -50mV (mV)	-7.57 ± 2.79	-8.06 ± 2.47	-7.78 ± 2.53	-10.5 ± 1.58

Tail current activation

The voltage dependence of the activation of I_{Ca} was determined independent of the changing driving force by tail current activation protocols. The tail currents were evoked by 5 ms voltage steps from E_{hold} to 40 mV in 5 mV increments (Fig. 3.4, A). The I/V relations were normalized to the maximal tail current of each cell and fit to first and second-order Boltzmann equations (Eq. 2.1) for E_{hold} -50 mV and E_{hold} -100 mV respectively (Fig. 3.4, B – C).

The Boltzmann fits revealed the voltage where halfmaximal activation occurs ($V_{0.5,act}$) and the slope factor (s_{act}). Compared to all other cohorts, $V_{0.5,act}$ was more depolarized in the 3w cohorts, for the total current as well as for HVA currents (Figs. 3.4, D1 & D2; 3.5, A & B). Differences in $V_{0.5,act}$ between all other cohorts were not statistically significant for total current and HVA currents.

Additionally, s_{act} was slower in the 3w cohort, compared to all other cohorts, again, for total current (Figs. 3.4, D1; 3.5, C) as well as HVA currents (Figs. 3.4, D2; 3.5, D). Taken together, these data suggest a shift in I/V relations of I_{Ca} during the transition from suckling to weaned mice.

Differences in s_{act} between all other cohorts were not statistically significant for total current and HVA currents.

The values of the tail current activation parameters for all cohorts are summarized in Tab. 3.2.

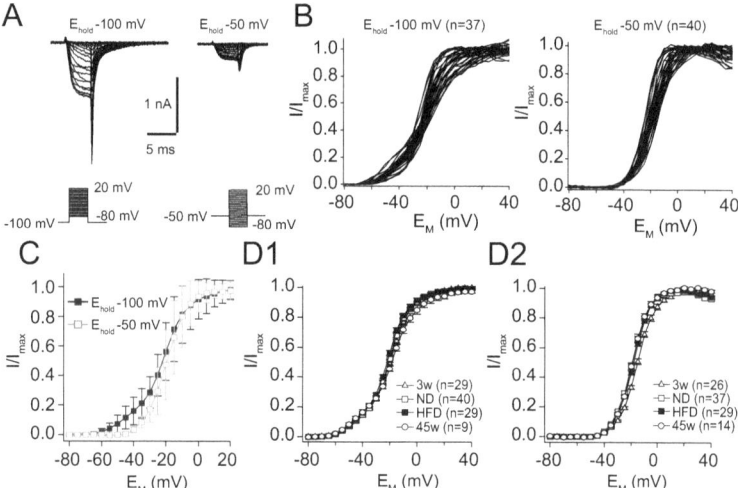

Figure 3.4: Tail current activation of I_{Ca}. (A) Example current traces for steady-state activation of tail currents from E_{hold} -100 mV and E_{hold} -50 mV respectively from the ND cohort. Tail currents were evoked by 5 ms voltage steps from E_{hold} to at least 20 mV in 5 mV increments. (B) I/V relations of tail currents normalized to the maximal tail current of each cell. Data from single neurons from the ND cohort with E_{hold} -100 mV (n=40) and -50 mV (n=37). (C) Mean I/V relations of steady-state tail current activation with E_{hold} -100 mV and E_{hold} -50 mV. Fits to second and first-order Boltzman relations (Eq. 2.1) respectively, revealed the following parameters: with E_{hold} -100 mV, voltage for half-maximal activation ($V_{0.5,act}$) = -33.06 ± 2.77 mV, slope factor (s_{act}) = 11.69 ± 3.1 (n=40); with E_{hold} -50 mV, $V_{0.5,act}$ = -19.36 ± 3.14 mV, s_{act} = 6.32 ± 1.17 (n=37). Error bars show SD. (D1) Mean I/V relations of steady-state tail current activation for all cohorts and E_{hold} -100 mV. Averaged data. Fits to second-order Boltzman relations revealed the following parameters: 3w, $V_{0.5,act}$ = -28.94 ± 3.02 mV, s_{act} = 13.81 ± 3.6; ND, $V_{0.5,act}$ = -33.06 ± 2.77 mV, s_{act} = 11.69 ± 3.1; HFD, $V_{0.5,act}$ = -31.29 ± 3.37 mV, s_{act} = 11.35 ± 3.05; 45w, $V_{0.5,act}$ = -31.58 ± 2.51 mV, s_{act} = 11.20 ± 2.41. (D2) Mean I/V relations of steady-state tail current activation for all cohorts and E_{hold} -50 mV. Averaged data. Fits to first-order Boltzman relations revealed the following parameters: 3w, $V_{0.5,act}$ = -15.79 ± 3.44 mV, s_{act} = 7.07 ± 0.91; ND, $V_{0.5,act}$ = -19.36 ± 3.14 mV, s_{act} = 6.32 ± 1.17; HFD, $V_{0.5,act}$ = -18.90 ± 2.39 mV, s_{act} = 6.27 ± 0.69; 45w, $V_{0.5,act}$ = -19.92 ± 2.33 mV, s_{act} = 6.11 ± 0.84. Error bars in D1 and D2 show SEM.

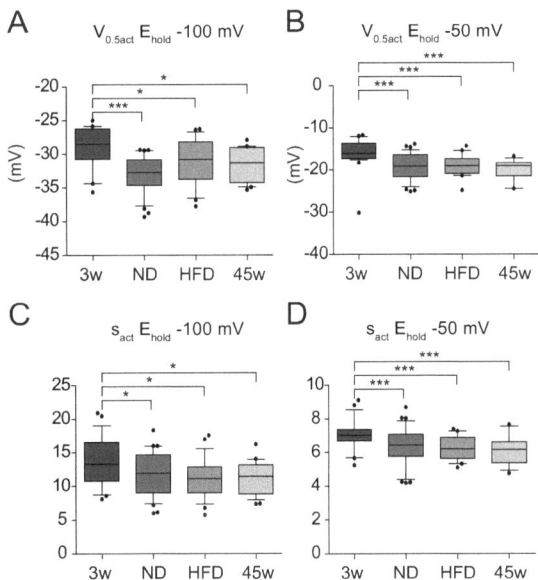

Figure 3.5: Tail current parameters from all cohorts. **(A)** Box plot of $V_{0.5,act}$ with E_{hold} -100 mV for all cohorts. Compared to all other cohorts $V_{0.5,act}$ is significantly more depolarized in the 3w cohort. **(B)** Box plot of $V_{0.5,act}$ with E_{hold} -50 mV for all cohorts. **(C)** Box plot of s_{act} with E_{hold} -100 mV for all cohorts. Compared to all other cohorts s_{act} is significantly smaller in the 3w cohort. Compared to all other cohorts $V_{0.5,act}$ is significantly more depolarized in the 3w cohort. **(D)** Box plot of s_{act} with E_{hold} -50 mV for all cohorts. Compared to all other cohorts s_{act} is significantly smaller in the 3w cohort. For details see Tab. 3.2 and Fig. 3.4, D1 and D2. Asterisks mark the level of significance: *** $p < 0.001$, * $p < 0.05$.

Table 3.2: Summary of steady-state tail current activation parameters of I_{Ca}.

	3w	ND	HFD	45w
$V_{0.5,act}$ E_{hold} -100mV (mV)	-28.94 ± 3.02 (n=29)	-33.06 ± 2.77 (n=40)	-31.29 ± 3.37 (n=29)	-31.58 ± 2.51 (n=9)
$V_{0.5,act}$ E_{hold} -50mV (mV)	-15.79 ± 3.44 (n=26)	-19.36 ± 3.14 (n=37)	-18.90 ± 2.39 (n=29)	-19.92 ± 2.33 (n=14)
s_{act} E_{hold} -100mV	13.81 ± 3.6	11.69 ± 3.15	11.35 ± 3.05	11.20 ± 2.41
s_{act} E_{hold} -50mV	7.07 ± 0.91	6.32 ± 1.17	6.27 ± 0.69	6.11 ± 0.84

3.2.2 Steady-state inactivation

Steady-state inactivation was measured from a holding potential of E_{hold} -100 mV. Calcium currents were elicited by a 5 ms test pulse to 0 mV. The test pulse was preceded by 500 ms pulses increasing from -95 to 0 mV in 5 mV increments (Fig. 3.6, A). The I/V relations were normalized to the maximal peak current of the test pulse for each cell. The steady-state inactivation of I_{Ca} started at prepulse potentials greater than -70 mV and increased with the amplitude of the depolarizing prepulse for all cohorts (Fig. 3.6, B & C). The normalized I/V relations were well fit with the sum of two Boltzmann equations (Eq. 2.2) indicating the existence of at least two inactivation processes (Fig. 3.6, C, D1 – E2). The steady-state inactivation parameters for all cohorts are summarized in Tab. 3.3.

Table 3.3: Summary of steady-state inactivation parameters of I_{Ca}.

	3w	ND	HFD	45w
$V_{1,0.5}$ (mV)	-39.52 ± 5.62 (n=23)	-36.32 ± 4.39 (n=24)	-36.98 ± 6.71 (n=20)	-36.31 ± 4.64 (n=11)
$V_{2,0.5}$ (mV)	-23.33 ± 3.08	-19.63 ± 7.7	-21.79 ± 4.62	-24.03 ± 4.52
s_1	20.29 ± 2.93	16.80 ± 5.36	17.34 ± 6.24	15.57 ± 4.07
s_2	4.48 ± 1.05	4.97 ± 2.84	4.47 ± 1.98	6.65 ± 2.93

3 Results

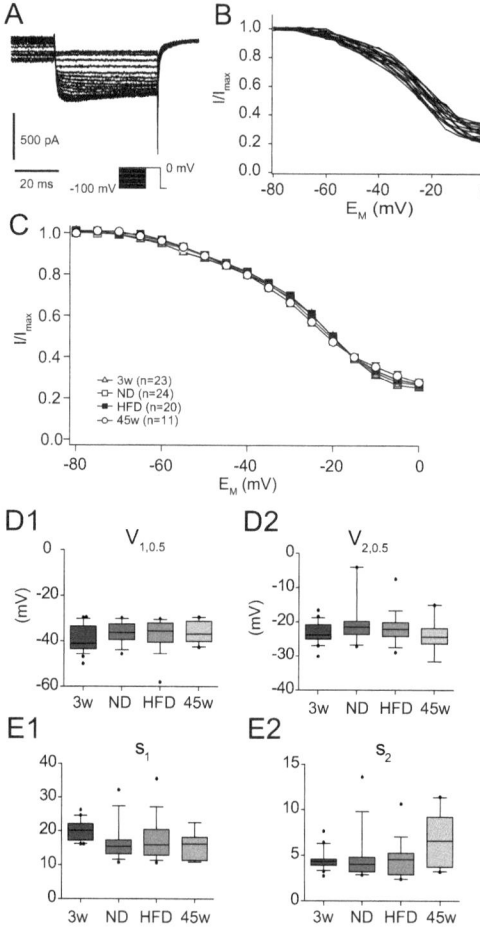

Figure 3.6: Steady-state inactivation of I_{Ca}. (A) Example current trace for steady-state inactivation of I_{Ca} from the ND cohort. The neurons were held at -100 mV. Calcium currents were elicited by a 5 ms test pulse to 0 mV. The test pulse was preceded by 500 ms pulses increasing from -95 to 0 mV in 5 mV increments. (B) I/V relations for steady-state inactivation of peak I_{Ca} normalized to the maximal current of each cell. Data from single neurons of the ND cohort (n=24). (C) Mean I/V relations of steady-state inactivation for all cohorts. The steady-state inactivation of I_{Ca} started at prepulse potentials greater than -70 mV for all cohorts. Averaged data. Error bars show SEM. Fits to the sum of two Boltzman equations (Eq. 2.2) revealed the following parameters: **(D1)** $V_{1,0.5}$: 3w, -39.52 ± 5.62 mV; ND, -36.32 ± 4.39 mV; HFD, -36.98 ± 6.71 mV; 45w, -36.31 ± 4.64 mV. **(D2)** $V_{2,0.5}$: 3w, -23.33 ± 3.08 mV; ND, -19.63 ± 7.7 mV; HFD, -21.79 ± 4.62 mV; 45w, -24.03 ± 4.52 mV. **(E1)** s_1: 3w, 20.29 ± 2.93 mV; ND, 16.80 ± 5.36 mV; HFD, 17.34 ± 6.24 mV; 45w, 15.57 ± 4.07 mV. **(E2)** s_2: 3w, 4.48 ± 1.05 mV; ND, 4.97 ± 2.84 mV; HFD, 4.47 ± 1.98 mV; 45w, 6.65 ± 2.93 mV. The differences in all parameters were not statistically significant.

3.2.3 Inactivation kinetics of the calcium current during a sustained pulse

To analyze the inactivation kinetics of I_{Ca}, long lasting depolarizing pulses were applied with E_{hold} -100 mV and E_{hold} -50 mV. The neurons were depolarized from E_{hold} to 0 mV for 1 s (Fig. 3.7, A). The currents were normalized to the maximal peak current of each cell and the decay of I_{Ca} for both holding potentials was well fit with a double exponential function (Eq. 2.3).

With E_{hold} -100 mV, the decay of I_{Ca} contained two different time constants, a fast ($\tau_1 E_{hold}$ -100 mV, e.g. 55.42 ± 13.70 ms in the ND cohort) and a slow time constant ($\tau_2 E_{hold}$ -100 mV, e.g. 852.1 ± 354.1 ms in the ND cohort, n=16). With E_{hold} -50 mV, the fit of the current decay revealed only two slow time constants ($\tau_1 E_{hold}$ -50 mV = 817.3 ± 580.1 ms; $\tau_2 E_{hold}$ -50 mV = 890.8 ± 558 ms; n=11) which were both in the same range as the slow time constant τ_2 with E_{hold} -100 mV (Fig. 3.7, B).

This indicates that the decay kinetics of the total calcium current in POMC neurons contain at least two inactivating components and that the fast inactivating component is inactivated at E_{hold} -50 mV. This suggests, that the fast inactivating component is mainly composed of (T-type-like) LVA currents and the remaining slow time constant belongs to HVA components of the calcium current.

Table 3.4: Summary of decay time constants of the inactivation kinetics of I_{Ca}.

	3w	ND	HFD	45w
Decay time constant τ_1 (ms)	43.66 ± 14.67 (n=23)	55.42 ± 13.7 (n=24)	51.28 ± 26.38 (n=20)	37.91 ± 20.34 (n=11)
Decay time constant τ_2 (ms)	868.7 ± 364.7	852.1 ± 354.1	1280 ± 620	1347 ± 1240

3 Results

The currents of the long depolarizing pulses for the overall current were normalized to the maximal peak current of each cell and the decay of I_{Ca} was fit with a double exponential function. Differences in the decay time constants of the fits were not statistically significant for both time constants between all cohorts. This indicates that there are no changes in the inactivation kinetics of I_{Ca} during a sustained depolarization. The time constants for all cohorts are summarized in Tab. 3.4.

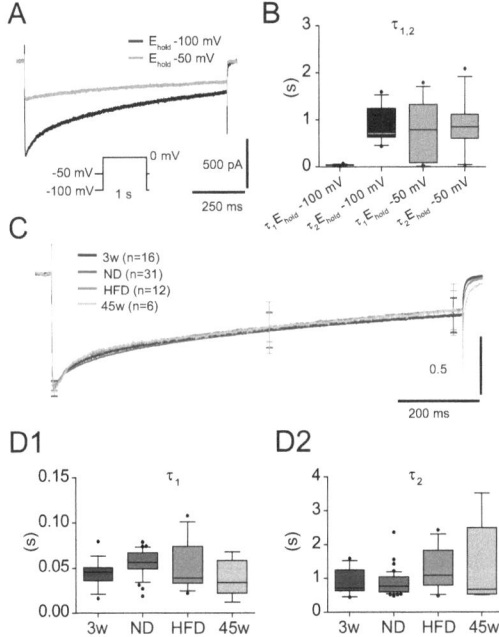

Figure 3.7: Decay time constants for inactivation kinetics of I_{Ca}. (A) Example current traces for inactivation kinetics of I_{Ca} from E_{hold} -100 mV and -50 mV respectively from the ND cohort. The neurons were depolarized from E_{hold} to 0 mV for 1 s. (B) The currents were normalized to the maximal peak current of each cell and the current decays were fit with a double exponential fit (Eq. 2.3). The fit of the decay of I_{Ca} from E_{hold} -100 mV contained a fast and a slow time constant (τ_1 = 55.42 ± 13.70 ms; τ_2 = 852.1 ± 354.1 ms; n=16) compared to the fit of the decay from E_{hold} -50 mV (τ_1 = 817.3 ± 580.1 ms; τ_2 = 890.8 ± 558 ms; n=11). This indicates that the decay kinetics of I_{Ca} contain at least two inactivating components. The fast inactivating component is inactivated at E_{hold} -50 mV suggesting that it is mainly composed of (T-type-like) LVA currents and that the remaining slowly inactivating component is mainly composed of HVA currents. (C) Mean current traces normalized to the maximal peak current of each cell from E_{hold} -100 mV for all cohorts. The decay of I_{Ca} was fit with a double exponential function (Eq. 2.3). Error bars show SD. (D1) Box plot of the fast time constants (τ_1) for all cohorts from the double exponential fit: 3w, 43.66 ± 14.67 ms; ND, 55.42 ± 13.70 ms; HFD, 51.28 ± 26.38 ms; 45w, 37.91 ± 20.34 ms. Differences in decay time constants between all cohorts were not statistically significant. (D2) Box plot of the slow time constants (τ_2) for all cohorts from the double exponential fit: 3w, 868.7 ± 364.7 ms; ND, 852.1 ± 354.1 ms; HFD, 1280 ± 620 ms; 45w, 1347 ± 1240 ms. Differences in decay time constants between all cohorts were not statistically significant.

3.3 Calcium handling in POMC neurons in the ARC

The cellular parameters that shape intracellular Ca^{2+} dynamics in POMC neurons in the ARC were analyzed for three different cohorts: P14 – 21 suckling milk (3w), P105 – 140 normal diet (ND) and P105 – 140 high-fat diet (HFD) mice. To determine the intracellular Ca^{2+} dynamics in POMC neurons the 'added buffer' approach in combination with whole-cell patch-clamp electrophysiology and ratiometric calcium imaging was used. The aim was to quantitatively analyze parameters that determine cytosolic Ca^{2+} dynamics in addition to voltage-activated Ca^{2+} influx: e.g. the endogenous Ca^{2+} binding ratio κ_s and the extrusion rate γ.

3.3.1 Calcium resting level

The Ca^{2+} resting levels were determined from the calibrated fura-2 fluorescence baseline before stimulation. The concentrations of free Ca^{2+} were determined from the ratio of the imaging signals according to Grynkiewicz *et al.* (1985, Eq. 2.4). The calibration was performed *in vitro* (in solution). For calibration, a drop of each calibration solution (75 µl, R_{min}: no Ca^{2+}, R_{def}: 0.35 µM Ca^{2+}, R_{max}: 10 mM Ca^{2+}) was placed on a sylgard coated recording chamber. Ratio images (from 340 and 380 nm excitation) were acquired for each solution. The acquired values were: correction factor $P = 0.93 \pm 0.10$, (Poenie, 1990); resulting in: $R_{max} = 1.738 \pm 0.003$, n=20; $R_{min} = 0.161 \pm 0.000$, n=20; $R_{def} = 0.641 \pm 0.002, n = 20$; $K_{d,fura,eff} = 0.798 \pm 0.003$ µM, the isocoefficient $\alpha = 0.160 \pm 0.020$ and $K_{d,fura} = 0.135 \pm 0.007$ µM (Eq. 2.6). The mean resting levels of free Ca^{2+} before the first stimulation for the different cohorts were: 3w, 0.032 ± 0.014 µM, n=17; ND, 0.021 ± 0.005 µM, n=10 and HFD, 0.040 ± 0.016 µM, n=12. The ND cohort displayed a lower resting Ca^{2+} concentration that was statistically significant compared to the 3w and HFD cohorts (Fig. 3.8, A).

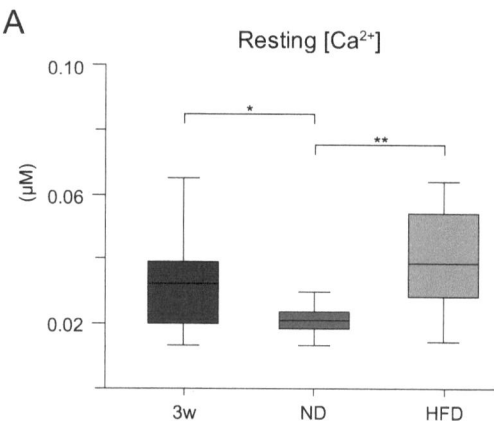

Figure 3.8: Calcium resting level and cell volume of POMC neurons. (A) Box plot of the resting Ca^{2+} concentration in POMC neurons. The Ca^{2+} resting levels were determined from the baseline of the fluorescence signal before stimulation. The values were as follows: 3w, 0.032 ± 0.014 µM, n=17; ND, 0.021 ± 0.005 µM, n=10 and HFD, 0.040 ± 0.016 µM, n=12. The difference in resting Ca^{2+} concentration of POMC neurons from the ND cohort compared to the 3w and HFD cohort were statistically significant. Asterisks mark the level of significance: ** $p < 0.01$, * $p < 0.05$.

3.3.2 Dye concentration from loading curves

In order to calculate the endogenous calcium handling parameters, Ca^{2+} transients at varying concentrations of the added buffer (fura-2) need to be recorded. Furthermore, it is crucial to know the concentration of the added buffer fura-2 at the time the transient calcium signals are recorded. To determine the intracellular concentration of fura-2 at any time during the experiment, a dye-loading curve was recorded before and after the Ca^{2+} transients were elicited. POMC neurons were loaded via the patch pipette containing 0.1 mM fura-2 and intracellular saline. Before break-in to whole-cell mode and during dye-loading, the fluorescence of the cell body was monitored with an excitation wavelength of 360 nm to determine the dye concentration and with an excitation wavelength of 380 nm to monitor the condition of the cell. Recordings where the neuron could not be properly loaded or where the condition of the cell deteriorated were discarded.

After establishing the whole-cell configuration, fura-2 was detected in the cell bodies within 30 seconds. During the loading of the cell, the fura-2 concentration was monitored and calcium transients were elicited at low, medium and high fura-2 concentrations with 50 ms pulses to 10 mV from a holding potential of -80 mV (Fig. 3.9, C & D). The fura-2 concentration increased until it reached a stable value, finally reaching the concentration of the pipette solution (0.1 mM). During the dye-loading, fluorescence images were obtained in 30 s intervals (Fig. 3.9, E). To determine the fura-2 concentration at the time the transients were recorded, the time course of the increasing fluorescence (loading curve) was fit with an exponential function for each experiment (Eq. 2.10, for details see Methods). The average time constants for dye loading determined from the exponential fits (Eq. 2.10) were as follows for the different cohorts: 3w, 1139 ± 840 s, ND n=18; 855 ± 554 s, n=12; HFD, 1065 ± 900 s, n=13. The differences in loading time constants were not statistically significant between the cohorts.

3 Results

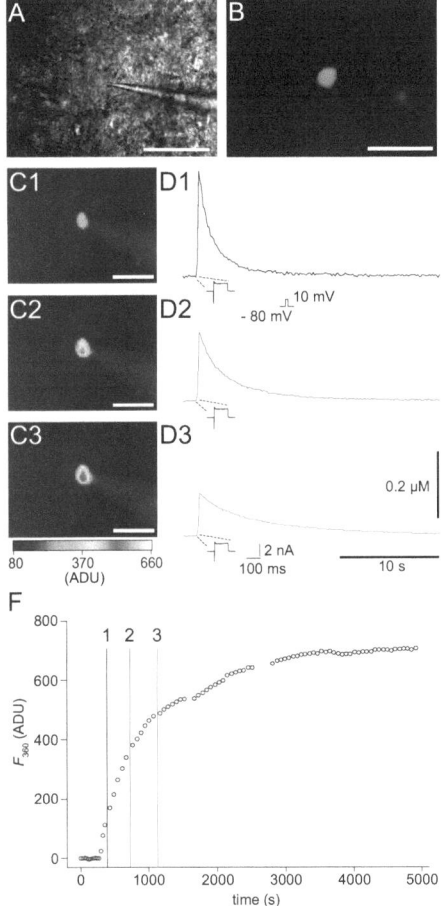

Figure 3.9: Fura-2 loading and decay kinetics. (A) Transmission image of the recording situation with the patch-clamp electrode on a POMC neuron in the living brain slice. **(B)** Fluorescence image showing the EGFP fluorescence signal of the POMC neuron. The area shown is the same as in A. **(C)** Sequence of fluorescence images showing the loading of fura-2 into the soma of a POMC neuron in the whole-cell configuration. The images were taken at 120 s (C1), 240 s (C2) and 480 s (C3) after break-in with an excitation wavelength of 360 nm (the isosbestic point of fura-2). Note the increase in fura-2 fluorescence over time, captured in analog-to-digital units (ADU) of the CCD chip. **(D)** Calcium transients of free Ca^{2+} concentrations at increasing fura-2 concentrations. Ca^{2+} transients were elicited with 50 ms pulses to 10 mV from a holding potential of -80 mV. The transients were elicited at the same timepoints as in C1 – C3, respectively. The recorded whole-cell current is shown below the trace. Note the decrease in signal amplitudes and increase in decay time constants. **(E)** Loading curve showing the increase in fura-2 concentration in a POMC neuron and the time points (1 – 3) the images and transients in C & D were recorded. The fura-2 concentration at the time of the transients was determined as described in Methods. Scale bar in A & B: 50 μm, in C 1 – 3: 20 μm.

3.3.3 Calcium handling properties

The kinetics of cytosolic Ca^{2+} signals are strongly dependent on endogenous and exogenous (added) Ca^{2+} buffers in the cell. The amplitude and decay rate of the free intracellular Ca^{2+} change with increasing concentration of the exogenous buffer. The amplitude of free Ca^{2+} decreases and the time constant of the decay ($\tau_{transient}$) increases (Figs. 1.4 & 3.9, C2 – E2). If the buffering capacity of the added buffer is known, $\tau_{transient}$ can be used to estimate the Ca^{2+} signal under conditions where only endogenous buffers are present ($-\kappa_B = 1 + \kappa_S$). The model used for this study (Eq. 2.9, Neher and Augustine, 1992) assumes that the decay time constants $\tau_{transient}$ are a linear function of the Ca^{2+} binding ratios (κ_B and κ_S). κ_S was determined from the negative x-axis intercept of the plot $\tau_{transient}$ vs. κ_B (Fig. 3.10). The point of intersection of the linear fit with the y-axis denotes the endogenous decay time constant τ_{endo} (no exogenous Ca^{2+} buffer in the cell). The slope of the fit is the inverse of the linear extrusion rate (γ). To estimate the variability of the parameters determined by linear fits, a bootstrap approach was used (n=1000) as described in the Methods chapter.

The endogenous decay time constants for the different cohorts were as follows: 3w, 1.2 ± 0.9 s; ND, 2.8 ± 0.9 s; HFD, 2.0 ± 0.5 s. The endogenous Ca^{2+} binding ratio for the different cohorts were as follows 3w, 84 ± 72; ND, 399 ± 221; HFD, 220 ± 106. The extrusion rates for each cohort were: 3w, 60 ± 14 s^{-1}; ND, 134 ± 39 s^{-1}; HFD, 107 ± 25 s^{-1}. The differences in endogenous Ca^{2+} binding ratios, endogenous decay time constants and extrusion rates were statistically significant between the three cohorts ($p < 0.001$, Fig. 3.11, A – C). These data suggest the following: the feeding of a high-fat diet decreased the endogenous buffer capacity and extrusion rate of POMC neurons. Interestingly, the high-fat diet also lead to a decrease in the endogenous decay time constant of the calcium signal. Additionally, in the transition from suckling to weaned mice (3w to ND), the decay time constant of the endogenous calcium signal increases in POMC neurons, the endogenous buffering capacity and the extrusion rate also increase. This trend is damped by the feeding of the high-fat diet, where POMC neurons have a slower decay time constant of the endogenous calcium signal, a lower endogenous buffering capacity and a slower extrusion rate than the normal diet feeding mice. The data is summarized in Tab. 3.5.

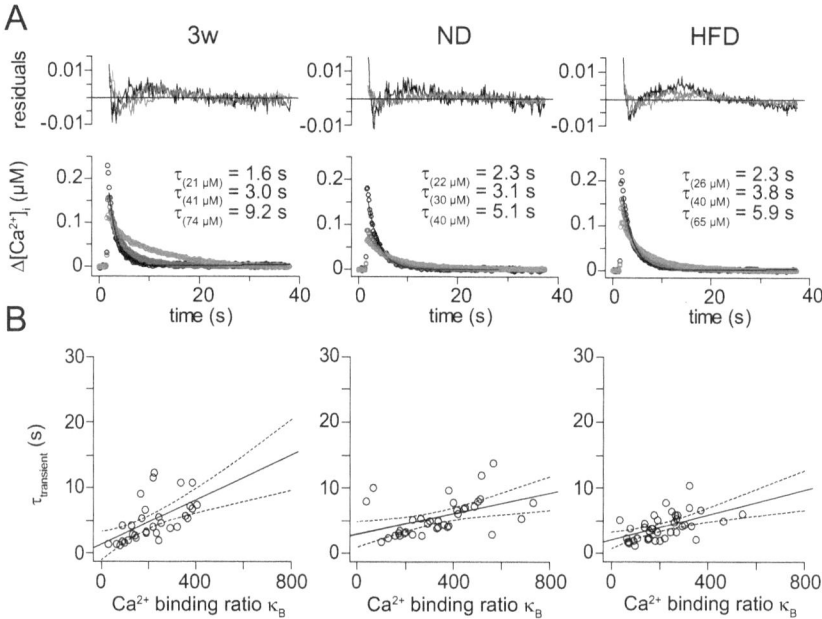

Figure 3.10: Calcium binding rates and calcium extrusion rates of POMC neurons in the ARC. (A) Calcium transients of the free Ca^{2+} concentration at increasing fura-2 concentrations from typical experiments of the 3w, ND and HFD cohorts. Ca^{2+} transients were elicited with 50 ms pulses to 10 mV from a holding potential of -80 mV. The decay time constant was well fit with a monoexponential function (Eq. 2.11, see residuals). The time constants of the transient (τ) for each cohort are given with the fura-2 concentration in subscripts at the time the transient was recorded. **(B)** The decay time constants of the Ca^{2+} transients were plotted as a function of the Ca^{2+} binding ratio (κ_B) which was calculated from the intracellular fura-2 concentration, the K_d of fura-2 and the resting concentration of free intracellular Ca^{2+} (see Methods). The data were fit with a linear function (Eq. 2.9) to determine the endogenous time constant for the decay of the Ca^{2+} signal (τ_{endo}, y-axis intercept), the endogenous Ca^{2+} binding ration (κ_S, x-axis intercept) and the extrusion rate (γ, slope of the fit). To estimate the variability of the linear fits, a bootstrap approach was used (n=1000) as decribed in Methods. Dashed lines the indicate 95 % confidence bands of the fit.

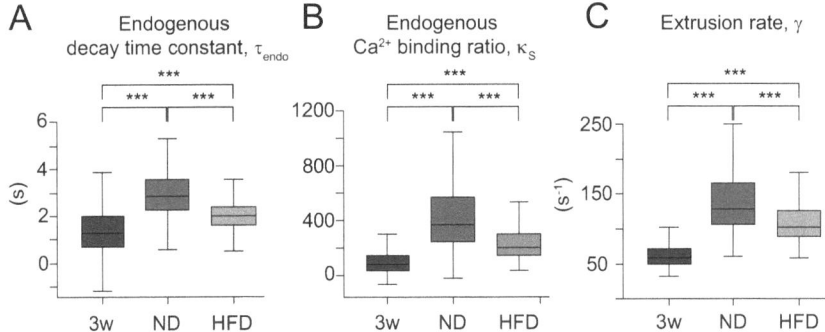

Figure 3.11: Calcium handling parameters for all cohorts. (A) Box plot showing the calculated values for the endogenous decay time constant (τ_{endo}). The endogenous decay time constants for the different cohorts were as follows: 3w, 1.2 ± 0.9 s; ND, 2.8 ± 0.9 s; HFD, 2.0 ± 0.5 s. **(B)** Box plot of the calculated values for the endogenous Ca^{2+} binding ratio (κ_S). The endogenous Ca^{2+} binding ratios for the different cohorts were as follows: 3w, 84 ± 72; ND, 399 ± 221; HFD, 220 ± 106. **(C)** Box plot showing the calculated values for the extrusion rate (γ). The extrusion rates for the different cohorts were as follows: 3w, 60 ± 14 s^{-1}; ND, 134 ± 39 s^{-1}; HFD, 107 ± 25 s^{-1}. The differences for each parameter between the cohorts were statistically significant. Asterisks mark the level of significance: *** $p < 0.001$.

Table 3.5: Summary of the calcium handling parameters of POMC neurons in the ARC.

	3w	ND	HFD
Resting [Ca^{2+}] (μM)	0.032 ± 0.014 (n=17)	0.021 ± 0.005 (n=10)	0.040 ± 0.016 (n=12)
Time constant of the dye loading (s)	1139 ± 840 (n=18)	855 ± 554 (n=12)	1065 ± 900 (n=13)
Endogenous decay time constant, τ_{endo} (s)	1.2 ± 0.9	2.8 ± 0.9	2.0 ± 0.5
Endogenous Ca^{2+} binding ratio, κ_S	84 ± 72	399 ± 221	220 ± 106
Extrusion rate, γ (s^{-1})	60 ± 14	134 ± 39	107 ± 25

3.4 POMC and SIM1 neurons in the PVH

The periventricular and medial parvocellular subdivisions of the PVH receive a dense innervation from POMC neuron fibers and also contain the MC4R (Fig. 3.12,A; Elmquist *et al.*, 1997; Bagnol *et al.*, 1999; Cowley *et al.*, 1999; Bouret *et al.*, 2004). Additionally, it is hypothesized that food intake is regulated in the PVH by an interaction between POMC neurons and putative second-order neurons to POMC neurons: SIM1 neurons (Balthasar *et al.*, 2005). The PVH has been studied previously in great detail and several neuron types could be characterized (Tasker and Dudek, 1991; Hoffman *et al.*, 1991; Tasker and Dudek, 1993; Luther and Tasker, 2000; Luther *et al.*, 2000; Stern, 2001; Luther *et al.*, 2002; Melnick *et al.*, 2007). To elucidate the characteristics of possible second-order neurons, SIM1 neurons were identified by EGFP fluorescence (Fig. 3.12, B). Because of the known innervation pattern of POMC neurons, the strict localization of the MC4R and the possible involvement in energy homeostasis and food intake regulation, the experiments were focussed on the periventricular and medial parvocellular subdivisions of the PVH (Swanson and Sawchenko, 1983). The SIM1 neuron population in the periventricular and medial parvocelluar part of PVH was studied in detail using whole-cell current clamp electrophysiology with the patch-clamp technique.

3 Results

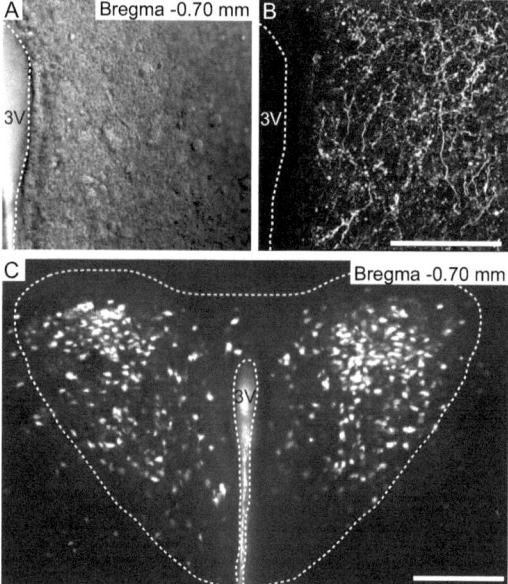

Figure 3.12: POMC and SIM1 neurons in the PVH. (A) Transmission image showing the right half of the periventricular and medial parvocellular PVH. The dashed line outlines the third ventricle. **(B)** Maximum intensity projection of a fluorescence image stack showing dense innervation of EGFP expressing fibers from POMC neurons in the PVH. The same area as shown in A. The images were acquired in the living slice using multiphoton microscopy. The dashed line outlines the third ventricle. **(C)** Merged fluorescence and transmission image of SIM1 neurons in the PVH in the living brain slice. SIM1 neurons expressing EGFP are distributed across the nucleus. The dashed line outlines the PVH. Scale bar in B: 100 µm, in C: 200 µm. 3V: third ventricle.

3.5 Comparison of cellular parameters from SIM1 labeled parvocellular neurons in the PVH

It was possible to identify previously described subtypes in the SIM1 expressing population of neurons in the periventricular (pv) and medial parvocellular (mp) subdivisions of the PVH (Tasker and Dudek, 1991; Tasker and Dudek, 1993; Luther and Tasker, 2000; Luther *et al.*, 2000; Stern, 2001; Luther *et al.*, 2002; Melnick *et al.*, 2007). These subtypes were parvocellular neurosecretory neurons (NS) and parvocellular pre-autonomic neurons (PA). In addition, SIM1 neurons recorded in the dorsal parvocellular and the magnocellular subdivisions of the PVH could be identified according to their electrophysiological profile as the bursting and magnocellular neurosecretory subtypes that have been previously found in the PVH (Figs. 3.13, 3.14; Tasker and Dudek, 1991). This demonstrates that SIM1 expressing neurons are a heterogenous population, spanning the known neuron subtypes in the PVH. However, because of the aforementioned focus on the pv and mp subdivisions, bursting and magnocellular SIM1 neurons were not further investigated. Electrophysiological properties of the NS and PA subtypes from the periventricular and medial parvocellular subdivision were investigated and are shown in Fig. 3.14. The neurons were recorded with whole-cell patch-clamp in current clamp mode. Current was injected for 500 ms from -100 pA to 40 pA in 10 pA steps. The NS subtype showed a weak spike frequency adaptation and a slow repolarization to E_M (Fig. 3.14, A1, A2). PA subtype neurons, however, showed a strong spike frequency adaptation and a relatively fast repolarization to E_M, followed by a low-threshold spike (Fig. 3.14, B1, B2). Both neuron types had a similar resting membrane potential, recorded right after the break-in to whole-cell mode (NS initial, -55.34 ± 7.3 mV; PA initial, -55.82 ± 6.03 mV; Fig. 3.14,

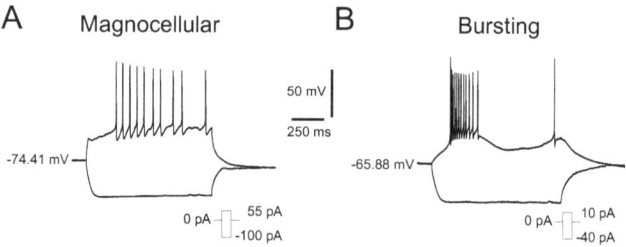

Figure 3.13: Magnocelluar neurosecretory and bursting subtypes of SIM1 neurons. (**A**) Example current clamp trace of a typical magnocellular SIM1 neuron recorded from the magnocellular subdivisions of the PVH. Single traces for 55 pA and -100 pA current injection. Note the characteristic delayed onset in firing. (**B**) Example current clamp trace of a typical bursting SIM1 neuron recorded from the dorsal region of the PVH. Single traces for 10 pA and -40 pA current injection. The neurons responded to positive current injection mainly with high-frequency bursts of action potentials riding on sometimes regenerative low-threshold spikes.

C). In both neuron types E_M hyperpolarized after break-in and became stable after 15 – 20 min. Differences between initial and stable E_M were statistically significant for each subtype (NS, $p < 0.05$; PA, $p < 0.01$). Differences between neuron types for initial and stable E_M, however, were not statistically significant (Fig. 3.14, C). The capacitance was compared between the two neuron types but the differences were also not statistically significant (NS, 12.83 ± 3.52 pF; PA, 13.56 ± 3.13 pF; Fig. 3.14, D). The time to repolarize after a hyperpolarization was analyzed by injecting hyperpolarizing current to change the membrane potential of the recorded neuron to -100 mV for 1 s. The differences were statistically significant between NS and PA neuron subtypes. It takes NS neurons significantly longer to reach E_M (NS, 0.278 ± 0.17 s; PA, 0.081 ± 0.06 s; $p < 0.001$, Fig. 3.14, E). The electrophysiological parameters of SIM1 NS and PA subtypes are summarized in Tab. 3.6.

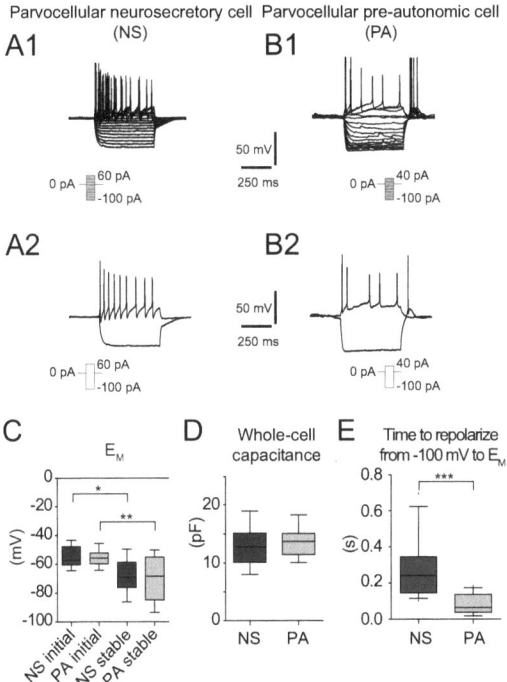

Figure 3.14: Electrophysiological properties of neurosecretory (NS) and pre-autonomic (PA) subtypes of SIM1 neurons. **(A1)** Example current clamp trace of a typical NS SIM1 neuron recorded from the periventricular (pv) and medial parvocellular (mp) subdivision of the PVH. Current was injected for 500 ms from -100 pA to 60 pA in 10 pA steps. **(A2)** Single traces for 60 pA and -100 pA current injection. Note the weak spike frequency adaptation and slow repolarization to resting membrane potential (E_M). **(B1)** Example current clamp trace for a typical PA SIM1 neuron recorded from the periventricular and medial parvocellular subdivision of the PVH. Current was injected for 500 ms from -100 pA to 40 pA in 10 pA steps. **(B2)** Single traces for 40 pA and -100 pA current injection. Note the strong spike frequency adaptation and relatively fast repolarization to E_M followed by a low-threshold spike (LTS). **(C)** Box plot of E_M for NS and PA subtypes immediately after break-in (initial) and after 20 min (stable). In both neuron types E_M hyperpolarized after break-in and became stable after 15 – 20 min. Values were as follows: NS initial, -55.34 ± 7.3 mV, n=15; NS stable, -69.3 ± 15.75 mV, n=15; PA initial, -55.82 ± 6.03 mV, n=15; PA stable, -67.61 ± 12.10 mV, n=11. Differences between initial and stable E_M were statistically significant for each subtype (NS, $p < 0.05$; PA, $p < 0.01$). Differences between neuron types for initial and stable E_M were not statistically significant. **(D)** Box plot of the whole-cell capacitance of NS and PA subtypes. Differences in capacitance were not statistically significant (NS, 12.83 ± 3.52 pF, n=15; PA, 13.56 ± 3.13 pF, n=15). **(E)** Box plot of the time it takes each subtype to reach E_M after a current injection to a membrane potential of -100 mV. It takes NS neurons significantly longer to reach E_M (NS, 0.278 ± 0.17 s, n=8; PA, 0.081 ± 0.06 s, n=7; $p < 0.001$). Asterisks mark the level of significance: *** $p < 0.001$, ** $p < 0.01$, * $p < 0.05$.

3.5.1 Spike frequency adaptation

The firing pattern of parvocellular NS and PA was analyzed. Current was injected for 500 ms from -100 pA to 60 pA in 10 pA steps (Fig. 3.15, A1, A2). Parvocellular NS neurons showed a more regular firing pattern with a weak spike frequency adaptation and more linear increase in frequency with increasing current injection (Fig. 3.15, A1). PA neurons showed a more irregular firing pattern with strong spike frequency adaptation and a less linear increase in firing frequency with increasing current injection (Fig. 3.15, A2). The instantaneous spike frequency (1/ISI) was calculated from a depolarizing current pulse (40 pA for 1 s) for both neuron types. The data were fit with a monoexponential function. NS neurons showed a weak spike frequency adaptation ($\tau = 226 \pm 176$ ms, n=6) compared to the PA subtype which exhibited a strong spike frequency adaptation ($\tau = 107 \pm 421$ ms, n=9). The spike frequency adaptation parameters are summarized in Tab. 3.6.

3.5.2 Slow afterhyperpolarization

To determine differences in the slow afterhyperpolarization between NS and PA subtypes, postitive current was injected (40 pA) for 1 s. Both SIM1 neuron subtypes displayed a slow afterhyperpolarization after a depolarizing current pulse was given that recovered back to E_M (Fig. 3.16, A). The differences in the time it took the neurons to repolarize back to E_M were not statistically significant (NS, 2.99 ± 2.88 s; PA, 0.8 ± 0.95 s). The maximal AHP for each neuron was calculated from the average over five current pulses. The values were as follows: NS, -6.9 ± 2.43 mV; PA -2.48 ± 2.27 mV (Fig. 3.16, B). NS subtypes had a significantly larger AHP than PA subtypes (NS, n=8; PA, n=7; $p < 0.01$). The AHP parameters are summarized in Tab. 3.6.

3 Results

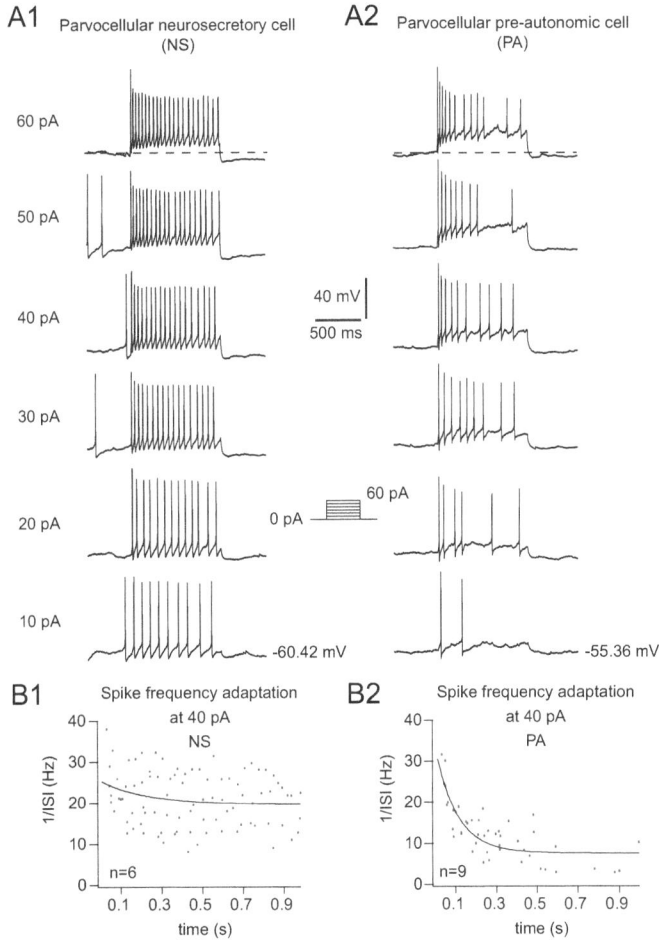

Figure 3.15: Difference in firing pattern between NS and PA subtypes. (A1) Example current clamp trace for the typical firing pattern of an NS neuron where positive current was injected from 10 to 60 pA for 500 ms. Note that there is a weak spike frequency adaptation and a slow afterhyperpolarization (AHP) after the current pulse (60 pA, dashed line). **(A2)** Example current clamp trace for the typical firing pattern of a PA neuron. Note that there is a strong spike frequency adaptation and a small AHP after the current pulse (60 pA, dashed line). **(B1)** Plot of the instantaneous spike frequency of a NS neurons during the injection of 40 pA. Each gray dot represents an interspike interval. The data were fit with a monoexponential function (Eq. 2.3). NS neurons showed a weak spike frequency adaptation ($\tau = 226 \pm 176$ ms, n=6). **(B2)** Plot of the instantaneous spike frequency of a NS neurons during the injection of 40 pA. Each gray dot represents an interspike interval. The data were fit with a monoexponential function. PA neurons showed a strong spike frequency adaptation ($\tau = 107 \pm 421$ ms, n=9).

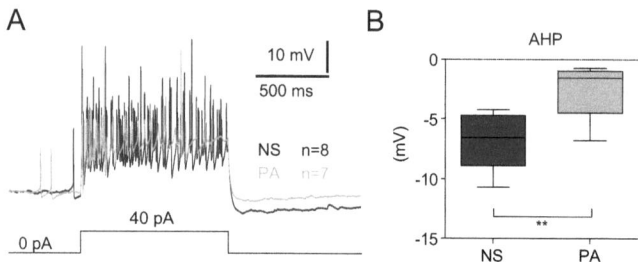

Figure 3.16: Slow afterhyperpolarization (AHP) in NS and PA subtypes of SIM1 neurons. (A) Averaged current clamp traces highlighting the differences in AHP after depolarzation for NS and PA subtypes. Positive current was injected (40 pA) for 1 s. Average from several neurons (NS, n=8; PA, n=7). **(B)** Box plot showing the maximal amplitude of the AHP for each subtype. The maximal AHP for each neuron was calculated from the average over five current pulses. The values were as follows: NS, -6.9 ± 2.43 mV; PA -2.48 ± 2.27 mV. NS subtypes had a significantly larger AHP than PA subtypes (NS, n=8; PA, n=7; $p < 0.01$).

3.5.3 Tolbutamide sensitivity

The adenosine triphosphate (ATP)-sensitive potassium (K_{ATP}) channel is widely distributed throughout the brain. It is assumed to play an important role in the signaling pathway of neuronal glucose sensitivity and possibly neuroprotection (Ashford *et al.*, 1990; Dunn-Meynell *et al.*, 1998; Miki *et al.*, 2001). Interestingly, in the PVH there is a non-homogenous distribution of neurons expressing the channel components (Dunn-Meynell *et al.*, 1998). Therefore, we adressed the question whether SIM1 neuron subtypes could be further classified by determining the sensitivity to the specific K_{ATP} channel blocker tolbutamide that would indicate a possible involvement in glucose sensing and energy homeostasis (Sakura *et al.*, 1995; Gribble *et al.*, 1997).

For these experiments, whole-cell current clamp recordings of SIM1 neurons were performed. When a stable E_M was reached after break-in (after $\sim 15-20$ min), 200 µM tolbutamide was applied to the neuron via the perfusion system for $\sim$15 min. In

63.63 % of the recorded SIM1 neurons the membrane potential depolarized during tolbutamide application and returned to control conditions after the wash-out of tolbutamide (Fig. 3.17,A & B). To see whether tolbutamide sensitivity could be correlated with the NS and PA neuron subtypes the percentages of tolbutamide sensitivity were calculated for each neuron type. Only 22.23 % of the recorded SIM1 NS subtypes were tolbutamide sensitive. Of the recorded PA subtypes, 92.31 % were tolbutamide sensitive. This suggests that the classified subtypes of NS and PA neurons each define heterogenous populations composed of tolbutamide sensitive and tolbutamide insensitive subgroups of neurons that might differentially express the K_{ATP} channel. The percentages of tolbutamide sensitivity are summarized in Tab. 3.6.

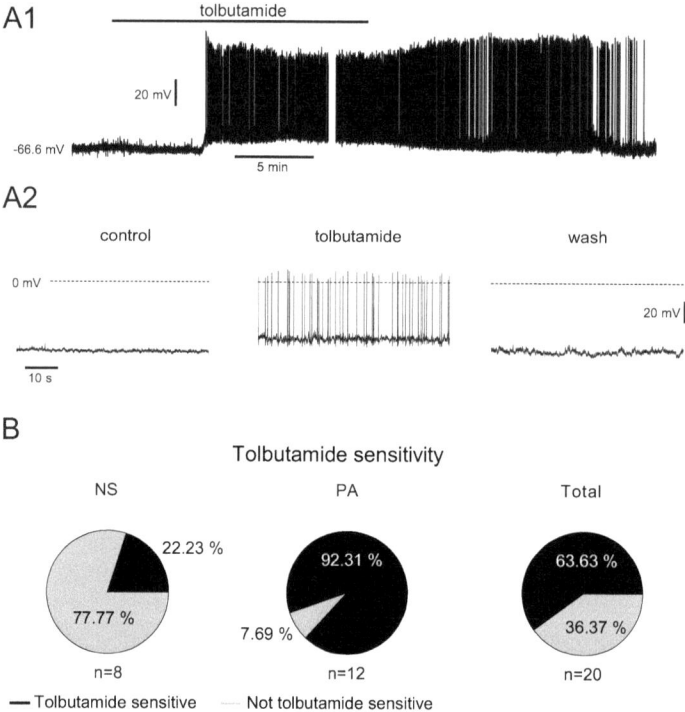

Figure 3.17: Tolbutamide sensitivity of SIM1 neurons. **(A1)** Representative current clamp trace of a SIM1 neuron where bath-application of the ATP-sensitive potassium (K_{ATP}) channel blocker tolbutamide (200 µM for ∼15 min) depolarizes the neuron above the action potential threshold. The effect is reversible after the wash-out of tolbutamide. **(A2)** Sections of the recording showing the membrane potential before (control), during (tolbutamide) and after (wash) tolbutamide application. **(B)** Percentages of tolbutamide sensitivity correlated to each neuron subtype. NS were only in 22.23 % of the cases sensitive to tolbutamide (n=8). PA neurons were in 92.31 % of the cases sensititve to tolbutamide (n=12). The overall sensitivity for tolbutamide of parvocellular SIM1 neurons in the periventricular and medial parvocellular subdivision of the PVH was 63.63 % (n=20).

Table 3.6: Summary of the electrophysiological parameters of NS and PA subtypes of SIM1 neurons.

	Parvocellular neurosecretory subtype (NS)	Parvocellular pre-autonomic subtype (PA)
E_M initial (mV)	-55.34 ± 7.3 (n=15)	-55.82 ± 6.03 (n=15)
E_M stable (mV)	-69.3 ± 15.75 (n=15)	-67.61 ± 7.3 (n=11)
Whole-cell Capacitance (pF)	12.83 ± 3.52 (n=15)	13.56 ± 3.13 (n=15)
Time to repolarize from -100 mV to E_M (s)	0.278 ± 0.17 (n=8)	0.08 ± 0.06 (n=7)
Decay time constant of the spike frequency adaptation, τ (ms)	226 ± 176 (n=6)	107 ± 421 (n=9)
Maximal amplitude of the afterhyperpolarization (mV)	-6.9 ± 2.43 (n=8)	-2.48 ± 2.27 (n=7)
Time to repolarize from afterhyperpolarization to E_M (s)	2.99 ± 2.88 (n=8)	0.8 ± 0.95 (n=7)
Sensitivity to tolbutamide (%)	22.23 (n=8)	92.31 (n=12)

4 Discussion

The aim of this study was to characterize and better understand the cellular parameters of neurons of the hypothalamic network that regulates energy homeostasis. This goal was approached in two parts.

First, the calcium handling parameters in first-order neurons of the hypothalamic energy balance network (POMC neurons) were analyzed. For this, voltage-activated calcium currents in POMC neurons were characterized by performing whole-cell patch-clamp voltage clamp recordings of identified POMC neurons in the living brain slice. The endogenous calcium binding ratio and extrusion rate were determined using the 'added buffer' approach. This was achieved by performing ratiometric calcium imaging experiments in combination with whole-cell patch-clamp recordings.

Second, to characterize potential candidates for second-order neurons to POMC neurons in the paraventricular nucleus of the hyptothalamus, SIM1 neurons were recorded with the whole-cell patch-clamp technique in the living brain slice.

4.1 POMC neurons

Neuronal calcium handling is known to change during development and aging (Fierro and Llano, 1996; Murchison and Griffith, 2007; Martella *et al.*, 2008; Foehring *et al.*,

2009). It is also known that the diet can modulate Ca^{2+} handling in neurons during aging (Hemond and Jaffe, 2005; Murchison and Griffith, 2007). To better understand the effect of a diet on Ca^{2+} handling in POMC neurons, two cohorts of P105 – 140 old mice that received either a high-fat or a normal chow diet were compared. It was found that the calcium current density and resting calcium concentration increased in POMC neurons when the animals were fed a high-fat diet. Furthermore, the parameters that shape the endogenous calcium signal (such as the endogenous calcium binding ratio and extrusion rate) changed under a high fat diet. In addition, experiments were performed in P14 – 20 suckling and P304 – 334 normal diet mice. These data were recorded to better understand age related changes in the Ca^{2+} handling of POMC neurons and, in addition, to lay the foundation for future studies. This will allow to gain an even clearer picture of the development of Ca^{2+} handling in POMC neurons during aging .

4.1.1 General properties

POMC neurons were recorded in the whole-cell configuration and stained with the biocytin/streptavidin technique. The morphology of POMC neurons was relatively heterogenous and could be grouped into two main categories: neurons with qualitatively complex branching patterns and neurons with a relatively simple branching pattern. In line with current data, these morphological data suggest that the POMC neuron population is heterogenous and that there are different POMC neuron subtypes fulfilling different functions. For example, it is known that some, but not all, POMC neurons express CART and cholinergic markers (Vrang *et al.*, 1999; Meister *et al.*, 2006) and that only a subpopulation of POMC neurons is sensitive to insulin (Ernst *et al.*, 2009). Very recently, a study has clearly shown, that there are two main subtypes that either release glutamate or GABA as their primary neurotransmitter (Hentges *et al.*, 2009). It is still not entirely clear if and how all these characteristics overlap and these are exciting issues to be explored in the future for a better characterization and subsequently, a better understanding of the characteristics and function of POMC neurons. Nevertheless, apparent differences in the electrophysiological responses of POMC neurons in voltage and current clamp could not be detected using our experimental paradigms, therefore data from POMC neurons of a single cohort were pooled.

4.2 Voltage-activated calcium currents in POMC neurons

4.2.1 Changes of calcium current parameters

One of the main findings of this thesis was that the P105 – 140 old mice that received a high-fat diet had a greater Ca^{2+} current density compared to the cohort of the same age that only received a normal diet (Fig. 3.3, A2, B2). This suggests that the high-fat diet has led to an increased Ca^{2+} influx in POMC neurons.

A possible explanation for this effect could be that factors secreted by adipose tissue (adipocytokines) directly or indirectly influence the expression or function of Ca^{2+} channels in POMC neurons. For example, increased leptin levels induced by an increased amount of adipose tissue could lead directly to the increased Ca^{2+} current density in the high-fat diet cohort (Frederich *et al.*, 1995). Leptin is known to increase the L-type Ca^{2+} current in POMC neurons (Wang *et al.*, 2008).

To test this, some factors need to be considered. First, the reported increase was observed during a short-term application to POMC neurons via the perfusion system in a primary cell culture of POMC neurons. Whether this change persists dur-

4 Discussion

ing prolonged exposure to high leptin levels needs to be determined. Second, diet-induced obese mice are known to become leptin resistant after 19 weeks (Lin et al., 2000). Therefore, it would be important to explore the development of this increase in current density in the weeks between P21 – P133 and whether current density continues to increase on the high-fat diet after week 19.

In addition, it cannot be excluded that other calcium channel types may also contribute to the increase in current density. Further experiments using channel specific blockers should reveal the contribution of each channel type to this increase in current density.

Other factors that are released by adipose tissue and that could possibly be involved in altering calcium handling would include: adiponectin, interleukin-6 (IL-6), tumor necrosis factor-α (TNF-α) and free fatty acids (FFA; Trayhurn and Beattie, 2001; Hauner, 2005; Rondinone, 2006). A recent study found that receptors for adiponectin are expressed in the ARC and colocalize with POMC (Guillod-Maximin et al., 2009). IL-6 is known to enhance Ca^{2+} responses of rat cerebellar granule neurons (Qiu et al., 1995). TNF-α is also known to influence neuronal and myocardial calcium handling (Motagally et al., 2009; Lee et al., 2007). FFA are known to modulate output of hypothalamic neurons (Foll et al., 2009).

These factors could also influence calcium currents indirectly. It is known that some of them are involved in the chronic inflammatory response that can be observed in obesity (Tilg and Moschen, 2006) and it is possible that an inflammatory response could influence calcium handling (Lu and Gold, 2008).

Another explanation for an increase in current density could be that an altered buffering capacity directly affects VGCCs. It was demonstrated that calcium buffering proteins can directly interact with VGCCs and alter the amplitude and inactivation kinetics of calcium currents (Meuth *et al.*, 2005; Lee *et al.*, 2006). Interestingly, the high-fat diet had no effect on the inactivation parameters of I_{Ca} (steady-state inactivation, inactivation kinetics during a sustained pulse; Figs. 3.6, C-E2 & 3.7, C-D2).

It was also found that there are slight changes in the current/voltage relations of I_{Ca} in POMC neurons during the transition from suckling to weaned mice and during 'early aging'. For example, in the 45w cohort (P304 – 334) compared to the ND cohort (P105 – 140), V_{max} of the HVA component was shifted to a more hyperpolarized potential. Other studies have also found that parameters of I_{Ca} can undergo changes already during 'early' aging. One study was able to show that the overall current increases in rat hippocampal neurons (Campbell *et al.*, 1996) and another found that there is a reorganization in the composition of HVA currents between P23 – P270 in mouse striatal neurons (Martella *et al.*, 2008). Other studies have found that in some regions of the brain, the current-voltage relations did not change at all during aging (Kostyuk *et al.*, 1993; Murchison and Griffith, 1995; Murchison and Griffith, 1996).

4.2.2 Methodical implications

Given the relatively complex morphology of POMC neurons (as shown in Fig. 3.1), proper voltage clamp is quite difficult to achieve in the living brain slice (Armstrong and Gilly, 1992; White *et al.*, 1995). In the recordings, however, blocking all currents that were not carried by Ca^{2+} made the neurons electrotonically more compact (Hille, 2001; Kloppenburg *et al.*, 2000). Recordings did not indicate poor voltage clamp (delay of current onset, jumps in the voltage dependence), suggesting that the recorded currents originated from well clamped regions of the neuron.

An important issue that needs to be considered for all whole-cell recordings is that in the whole-cell configuration, the cytosol is controlled by the pipette solution. On one hand, this is can be an advantage where recordings with a defined cytosolic concentration are desirable (for example, to define the intracellular fura-2 concentration during the 'added buffer' approach experiments). On the other hand, it needs to be kept in mind that the whole-cell configuration can lead to a wash-out of molecules for important cell functions (Pusch and Neher, 1988). For example, a small calcium current run-down, possibly due to wash-out, was noticeable during the calcium current recordings. However, the experimental protocols were started quickly after break-in to minimize the effect of a wash-out.

The determined parameters of I_{Ca} were well within range of previously reported calcium currents in vertebrate neurons (Avery and Johnston, 1996; Murchison and Griffith, 1996; Kammermeier and Jones, 1997; Elsen and Ramirez, 1998; Chambard *et al.*, 1999; Shirasaka *et al.*, 2004; Tanaka *et al.*, 2007; Zhang *et al.*, 2007; Wang *et al.*, 2008). The current densities determined for POMC neurons in this study (between -53.95 and -69.50 pA/pF for the total current and between -29.79 and -43.59 pA/pF for the HVA currents) seem to be in the same range with data from another study (Wang *et al.*, 2008; approx. 40 pA/pF for HVA currents, assuming an average whole-cell capacitance of 15 pF). In the same study, however, V_{max} was shifted more positively for the HVA currents (10 mV; Wang *et al.*, 2008). This could be explained by considering that most of the studies of I_{Ca} in hypothalamic neurons were done with acutely dissociated neurons. Differences in I_{Ca} parameters could be attributed to the circumstance that neurons in the living brain slice might possess a slightly different calcium current profile (e.g. because of an intact neurite structure), compared to neurons that underwent the dissociation and culturing process.

When pooling the currents from POMC neurons of a single cohort (for example for the ND cohort; Fig. 3.2, C), the I/V curves for the neurons were distributed evenly.

4 Discussion

This would suggest that the POMC neuron population, despite its heterogenous composition, exhibits a rather homogenous distribution of Ca^{2+} current parameters.

4.3 Intracellular calcium handling in POMC neurons

Important parameters for intracellular calcium handling (the calcium resting level, endogenous decay time constant of the calcium signal, the endogenous calcium binding ratio and the calcium extrusion rate) of POMC neurons were determined with the 'added buffer' approach for three different mice cohorts: P14 – 21, milk suckling (3w), P105 – 140, normal diet (ND) and P105 – 140, high-fat diet (HFD). This was achieved by performing whole-cell patch-clamp recordings in combination with ratiometric calcium imaging experiments of identified POMC neurons selectively expressing EGFP in the living brain slice.

4.3.1 Changes in intracellular calcium handling

The calcium handling parameters in POMC neurons of the mice cohorts that either received a high-fat or a normal diet were compared. The most important finding was that the high-fat diet had an effect on the calcium handling of POMC neurons. For example, the endogenous calcium binding ratio of the cohort that received the high-fat diet was almost halved compared to mice of the same age that received the normal chow diet (Fig. 3.11, B). Furthermore, the high-fat diet had also an effect on the resting Ca^{2+} resting level. It was almost twice as high in POMC neurons of animals from the HFD cohort compared to the normal diet cohort (Fig. 3.8). Interestingly, the extrusion rate in POMC neurons of the HFD cohort was lower but the determined endogenous time constant of the calcium signal was longer compared to the normal diet cohort.

This could be explained by the lower buffering capacity in the HFD cohort (Fig. 3.11). The lower buffering capacity would shorten the calcium signal as a greater amount of free calcium is available to be removed from the cytosol.

What does all this mean for POMC neurons? Proposing physiological consequences for one changing Ca^{2+} handling parameter (Ca^{2+} influx, Ca^{2+} buffering, Ca^{2+} extrusion) at a time is relatively straightforward. Understanding the combined effects of changes in all three parameters, however, is not as trivial. Nevertheless, a few points can be discussed.

The elevated resting Ca^{2+} levels, for example, could lead to an increased activation of calcium-dependent potassium (K(Ca)) channels, thus leading to a hyperpolarization and silencing of POMC neurons. It could be shown in recent experiments in our laboratory that POMC neurons recorded from the HFD cohort are more hyperpolarized and a greater percentage of POMC neurons is silenced compared to POMC neurons of the ND cohort (personal communication Simon Heß). This would mean that the observed effects of a high-fat diet could lead to a silencing of POMC neurons, a possible reduction in satiety signaling and further, increased food intake and reduced energy expenditure. Additional experiments to determine the contribution of K(Ca) channels to this silencing are currently performed in our laboratory.

A lower buffering capacity in combination with a greater calcium influx via VGCCs in POMC neurons of the HFD cohort would lead to a greater amount of free Ca^{2+} during calcium transients. In addition to the increased resting calcium level, this would increase the Ca^{2+} load on POMC neurons over time. Considering the aforementioned 'calcium hypothesis' of neuronal aging, it could be hypothesized that in the long run, the Ca^{2+} load on POMC neurons would increase and could also lead to a decline in POMC neuron function, synaptic release and synaptic plasticity. This would be analogous to the changes in calcium handling in hippocampal and cortical neurons that are thought to be responsible for the cognitive decline during aging. In the case of POMC neurons this would lead to altered signaling of satiety and changes in the energy uptake behaviour of the organism.

Possible reasons for the observed changes in calcium handling parameters could be, as mentioned above, factors secreted by the adipose tissue (e.g. adiponectin, IL-6, TNF-α and FFA) directly or indirectly acting on POMC neurons. In addition, these changes could also be a compensatory reaction to the increased Ca^{2+} influx. To which extent these factors are relevant, still needs to be explored by further experiments.

Additionally, the calcium handling for P14 – 21 mice was characterized. It was found that the calcium handling parameters change during the transition from suckling to weaned mice. For example, the buffering capacity increased from P14 – 21 to P105 – 140 old mice. This is in line with observations from rat cerebellar Purkinje neurons and dopaminergic rat substantia nigra neurons where the buffering capacity also increased in animals aged from P6 to P15 and from P13 - 17 to P25 - 32 (Fierro and Llano, 1996; Foehring *et al.*, 2009).

4.3.2 Individual parameters

Resting calcium level

The resting calcium levels determined for POMC neurons in this study ranged between 0.021 (for the ND cohort) and 0.040 µM (for the HFD cohort). This is well within the range of previously observed levels in vertebrate neurons (0.015 – 0.022 µM for proximal dendrites in hippocampal neurons, Liao and Lien, 2009; 0.046 µM in the rat calyx of Held, Helmchen *et al.*, 1997; 0.071 µM in rat fast-spiking hippocampal basket cells, Aponte *et al.*, 2008; 0.145 – 0.260 µM in rat neurohypophysial nerve endings, Stuenkel, 1994; 0.157 µM in mouse motoneurons of the nucleus hypoglossus, Lips and Keller, 1998).

Calcium buffering

The determined endogenous buffering capacity (κ_s) for POMC neurons ranged between 84 (for the 3w cohort) and 399 (for the ND cohort). This would would mean that during a Ca^{2+} transient, between 1.21 % and 2.5 % of the Ca^{2+} ions entering the cytosol would remain free. This is in line with buffering capacities observed in other vertebrate neurons. Previous studies found estimates for κ_s that range from 20 - 40 (inhibitory hippocampal interneurons, Foehring *et al.*, 2009; adrenal chromaffin cell, Zhou and Neher, 1993; calyx of Held, Helmchen *et al.*, 1997; mouse hypoglossal motoneurons, Lips and Keller, 1998) over 100 - 174 (cortical layer V neurons, Helmchen *et al.*, 1996; neurohypophysial nerve endings, Stuenkel, 1994;) to 900 - 2000 (rat superior cervical ganglion neurons, Wanaverbecq *et al.*, 2003; rat cerebellar Purkinje neurons, Fierro and Llano, 1996).

Calcium extrusion and endogenous decay constants

The endogenous decay time constants of the calcium signal in POMC neurons were extrapolated to conditions where no exogenous buffer is present. For POMC neurons they were found to be between 1.2 s (3w) and 2.8 s (ND) and are well within the range that has been reported previously. Previous estimates for decay time constants of the calcium signal ranged between 0.7 s in mouse hypoglossal motoneurons (Lips and Keller, 1998), 1.7 s in mouse oculomotor neurons (Vanselow and Keller, 2000), 3 s in rat Purkinje cells (Fierro *et al.*, 1998) and 1 - 5 s in rat nucleus basalis neurons (Tatsumi and Katayama, 1993). The estimates for the calcium extrusion rate in POMC neurons was between 60 s^{-1} (3w) and 134 s^{-1} (ND). This is in accord with other studies that have found extrusion rates between 60 s^{-1} in mouse hypoglossal neurons (Lips and Keller, 1998), 156 s^{-1} in mouse oculomotor neurons (Vanselow and Keller, 2000), 580 s^{-1} in dendrites of fast spiking rat hippocampal basket cells (Aponte *et al.*, 2008), 900 s^{-1} in the rat calyx of Held (Helmchen *et al.*, 1997) and 2000 s^{-1} in dendritic regions of rat cortical layer V neurons (Helmchen *et al.*, 1996).

4.3.3 Methodical implications

To obtain meaningful values using the 'added buffer' approach, the fluorimetric calcium indicator needs to be calibrated to determine the relationship between fluorescence ratio and calcium concentration. This was done according to Grynkiewicz *et al.* (1985). The data acquisition and calibration process, however, can introduce several sources for error that need to be considered. First, the preparation and imaging setup themselves can introduce errors during the acquisition of the fura-2 fluorescence (background fluorescence, bleaching, errors during detection, amplification and digitization of the fluorescence; Moore *et al.*, 1990). To keep these errors minimal, the experimental procedure was optimized to obtain fluorescence signals with a resonable signal to noise ratio while keeping the exposure time, excitation intensity and thus the bleaching as low as possible. Additionally, we applied a background reduction protocol (see Methods). Second, errors can occur during the calibration process of fura-2. A source of error could be, for example, a miscalculation of the actual free Ca^{2+} concentration in the calibration solutions due to pipetting errors, impurities and variations in temperature and pH (McGuigan *et al.*, 2007). Additional errors could be introduced by differences in fura-2 calibration parameters (e.g. K_d) due to differences in viscosity between the calibration solution and the cytosol or due to fura-2 decomposition (Moore *et al.*, 1990; Poenie, 1990; personal communication, Andreas Pippow). To ensure a precise estimate of the free Ca^{2+} concentration during the fura-2 calibrations for this thesis, the calibration solutions were freshly made and the free Ca^{2+} concentration of calibration solutions was determined with a Ca^{2+} selective electrode (see Methods; McGuigan *et al.*, 1991).

4.3.4 Outlook

It has to be considered that in addition to the reported findings, this study has also established a basis for further exploration of the larger complex of changes in calcium homeostasis that are induced by diet and happen during aging. Additional experiments should be performed to further elaborate the findings of this thesis. Several resulting questions should be addressed:

Are the observed changes in Ca^{2+} homeostasis restricted to POMC neurons or is this a general effect in hypothalamic neurons? To address this question experiments could be performed, for example, where the same experiments as in this thesis are performed in brain slices of the same mouse line but with neurons of the ARC that do not express EGFP (non-POMC neurons).

Are these changes in the Ca^{2+} homeostasis of POMC neurons reversible? To address this question, the Ca^{2+} homeostasis in POMC neurons of HFD mice that were switched to a normal chow diet at P140 could be analyzed at P315 and compared to the data from normal chow diet mice.

Does caloric restriction have an effect on the Ca^{2+} homeostasis of POMC neurons? In this case, a third cohort of mice that would have received a low calorie diet from P21 on could be compared to the data from ND and HFD mice recorded during this

4 Discussion

thesis.

As mentioned earlier, it would also be of interest to determine the contribution of the different calcium channel types to the observed changes in current density and I/V relations. For this, calcium current recordings with channel specific blockers could be performed.

To gain a greater understanding of the complex changes in the parameters that shape intracellular Ca^{2+} dynamics found during this thesis, simulations of the intracellular free Ca^{2+} dynamics using the determined parameters (calcium influx, calcium binding ratio, calcium extrusion rate) for POMC neurons should be performed (Pippow et al., 2009). This would help to illustrate and better understand the impact of the observed changes on the endogenous intracellular Ca^{2+} dynamics of POMC neurons.

Furthermore, the contribution of buffering proteins to the calcium binding ratio in POMC neurons would also grant a better understanding of the observed the changes in Ca^{2+} homeostasis. To determine the contribution of involved buffering proteins, quantitative single-cell reverse transcription polymerase chain reaction (RT-PCR) could be performed subsequent to whole-cell patch-clamp experiments (Liss and Roeper, 2004).

It should also be investigated how different Ca^{2+} clearance systems contribute to the extrusion rate (Fierro et al., 1998). For example, experiments are currently being performed in our laboratory to assess the contribution of mitochondrial Ca^{2+} clearance and the possible involvement of reactive oxygen species as a cause for the reduced extrusion rate in POMC neurons of the HFD cohort.

4.4 SIM1 neurons

It is known that the periventricular and medial parvocellular subdivisions of the PVH are POMC neuron target regions with dense innervation by POMC neurons and that SIM1 neurons play an important role in melanocortin mediated satiety signaling (Cowley *et al.*, 1999; Balthasar *et al.*, 2005). Previous studies have also revealed that the PVH contains electrophysiologically distinct neuron types (Tasker and Dudek, 1991; Luther and Tasker, 2000; Luther *et al.*, 2002). However, the electrophysiological parameters of SIM1 neurons (a genetically defined subpopulation of PVH neurons) have not yet been elucidated. Therefore, the electrophysiological responses of SIM1 neurons to current injection were characterized. It was found that SIM1 neurons in the PVH comprise a heterogenous population of electrophysiologically and pharmacologically distinct neuron subtypes. In addition, the found SIM1 neuron subtypes could be matched to neuron types that have been described previously in the PVH. Further, two subtypes of parvocellular SIM1 neurons (pre-autonomic and neurosecretory neurons) that could be candidates for second-order neurons to POMC neurons were characterized by their active membrane properties and sensitivity to the ATP-sensitive potassium channel blocker tolbutamide.

4.4.1 Parvocellular SIM1 subtypes in the pvPVH and mpPVH

It was found that SIM1 neurons of the mp and pv subdivisions can be classified into the two known parvocellular sub-types: neurosecretory and pre-autonomic neurons. Parvocellular neurosecretory neurons displayed a weaker spike frequency adaptation, a larger afterhyperpolarization after depolarizing current injection and it took longer for them to reach E_M after a hyperpolarization to -100 mV compared to pre-

4 Discussion

autonomic neurons. The neuron sub-types further showed different sensitivities to the K_{ATP} channel specific blocker tolbutamide. Only a minority of recorded parvocellular neurosecretory SIM1 neuron subtypes reacted to tolbutamide application with a depolarization and increase in action potential firing frequency that was reversible during wash-out (22.23 %). Of the recorded pre-autonomic subtype, in contrast, the majority was sensitive to tolbutamide (92.31 %). This suggests that both, the NS and the PA subtype of SIM1 neurons, comprise further subgroups of neurons. This is in line with an electrophysiological and morphological study where PA neurons in the PVH were identified by retrograde tracing and could be classified into different subgroups (Stern, 2001).

To further characterize the different subgroups of parvocellular neurons in the PVH, the underlying currents for the observed responses to current injection should be determined by performing whole-cell voltage clamp experiments where the specific currents can be isolated by ion substitution and specific blockers.

For example, characteristics and possible differences in the potassium A-current in NS neurons should be further analyzed. The A-current could be responsible for the delay in repolarization to E_M observed in NS neurons (Connor and Stevens, 1971; Rush and Rinzel, 1995). Further, the parameter space of the Ca^{2+}-dependent potassium current $I_{K(Ca)}$ should be explored as K(Ca) channels are known to be involved in the slow AHP following depolarization and spike frequency adaptation (Yarom *et al.*, 1985).

The hyperpolarization of the membrane potential that was observed in the whole-cell recordings of SIM1 neurons could be due to wash-out of ATP. In the absence of ATP, the K_{ATP} channels would open and thus hyperpolarize the neuron. However, the ATP concentration in the pipette solution (3 mM) should be high enough to keep

4 Discussion

the channels closed. It was previously reported that an ATP concentration of 1 mM is able to almost completely inhibit K_{ATP} channels (Gribble *et al.*, 1997). Also, if the K_{ATP} channels were exclusively responsible for the hyperpolarization in whole-cell mode, PA neurons would possibly hyperpolarize less, considering the observed differences in tolbutamide sensitivity and hypothesized differences in K_{ATP} channel expression. As this is not the case, the wash-out of other second messengers might also contribute to the hyperpolarization of the resting membrane potential after break-in. To avoid this issue, perforated patch recordings should be performed for current clamp recordings as they leave the intracellular second messenger concentration mostly intact (Horn and Marty, 1988; Akaike and Harata, 1994).

To gain a better understanding of the morphology and distinguish projection patterns of possible neurosecretory neuron subtypes, stainings should be performed in sagittal brain slices to keep the projection towards the median eminence intact.

Lastly, experiments with MC4R agonists (such as melanotan II; Fan *et al.*, 1997) should be performed, to determine which SIM1 neuron subtypes in the PVH are directly involved in energy homeostasis as second-order neurons to POMC neurons. However, to perform these experiments, a greater understanding of the parvocellular subtypes might be needed. To be able to distinguish between possible subpopulations of PA and NS subtypes could be crucial for these experiments as it is possible that only a small subpopulation of PA or NS neurons actually expresses the MC4R.

List of Tables

3.1	Summary of steady-state activation parameters of I_{Ca}.	56
3.2	Summary of steady-state tail current activation parameters of I_{Ca}.	60
3.3	Summary of steady-state inactivation parameters of I_{Ca}.	61
3.4	Summary of decay time constants of the inactivation kinetics of I_{Ca}.	63
3.5	Summary of the calcium handling parameters of POMC neurons in the ARC.	76
3.6	Summary of the electrophysiological parameters of NS and PA subtypes of SIM1 neurons.	87

List of Figures

1.1	Diagram of the neurocentric model of energy homeostasis	18
1.2	Hypothalamic nuclei involved in energy homeostasis and location of POMC neurons	19
1.3	Output projections from the PVH and the corresponding neuron types	22
1.4	Single compartment model	25
2.1	Background fluorescence subtraction	37
2.2	Estimation of the isocoefficient α	40
2.3	Fitting the loading curves	45
3.1	Diverse morphology of POMC neurons	51
3.2	I_{Ca} in POMC neurons	54
3.3	I_{Ca} current density and I/V relation in different age and diet cohorts.	55
3.4	Tail current activation of I_{Ca}	58
3.5	Tail current parameters from all cohorts	59
3.6	Steady-state inactivation of I_{Ca}	62
3.7	Decay time constants for inactivation kinetics of I_{Ca}	65
3.8	Calcium resting level and cell volume of POMC neurons	68
3.9	Fura-2 loading and decay kinetics	71

List of Figures

3.10 **Calcium binding rates and calcium extrusion rates of POMC neurons in the ARC** 74

3.11 **Calcium handling parameters for all cohorts** 75

3.12 **POMC and SIM1 neurons in the PVH** 78

3.13 **Magnocelluar and bursting subtypes of SIM1 neurons** 80

3.14 Electrophysiological properties of NS and PA subtypes of SIM1 neurons 81

3.15 **Difference in firing pattern between NS and PA subtypes** 83

3.16 **AHP in NS and PA subtypes of SIM1 neurons** 84

3.17 **Tolbutamide sensitivity of SIM1 neurons** 86

References

Akaike N, Harata N (1994) Nystatin perforated patch recording and its applications to analyses of intracellular mechanisms. *The Japanese Journal of Physiology* 44:433–473.

Aponte Y, Bischofberger J, Jonas P (2008) Efficient Ca2+ buffering in fast-spiking basket cells of rat hippocampus. *The Journal of Physiology* 586:2061–2075.

Archer ZA, Mercer JG (2007) Brain responses to obesogenic diets and diet-induced obesity. *The Proceedings of the Nutrition Society* 66:124–30.

Armstrong CM, Bezanilla F (1974) Charge movement associated with the opening and closing of the activation gates of the Na channels. *J Gen Physiol* 63:533–52.

Armstrong CM, Gilly WF (1992) Access resistance and space clamp problems associated with whole-cell patch clamping. *Methods in Enzymology* 207:100–122.

Arora S, Anubhuti (2006) Role of neuropeptides in appetite regulation and obesity–a review. *Neuropeptides* 40:375–401.

Ashford ML, Boden PR, Treherne JM (1990) Glucose-induced excitation of hypothalamic neurones is mediated by ATP-sensitive K+ channels. *Pflügers Archiv: European Journal of Physiology* 415:479–483.

Augustine GJ, Santamaria F, Tanaka K (2003) Local calcium signaling in neurons. *Neuron* 40:331–46.

Avery RB, Johnston D (1996) Multiple channel types contribute to the low-voltage-activated calcium current in hippocampal CA3 pyramidal neurons. *The Journal of Neuroscience* 16:5567–5582.

Bagnol D, Lu XY, Kaelin CB, Day HE, Ollmann M, Gantz I, Akil H, Barsh GS, Watson SJ (1999) Anatomy of an endogenous antagonist: relationship between Agouti-related protein and proopiomelanocortin in brain. *The Journal of Neuroscience* 19:RC26.

Balthasar N, Dalgaard LT, Lee CE, Yu J, Funahashi H, Williams T, Ferreira M, Tang V, McGovern RA, Kenny CD, Christiansen LM, Edelstein E, Choi B, Boss O, Aschkenasi C, yu Zhang C, Mountjoy K, Kishi T, Elmquist JK, Lowell BB (2005) Divergence of melanocortin pathways in the control of food intake and energy expenditure. *Cell* 123:493–505.

Benoit SC, Air EL, Coolen LM, Strauss R, Jackman A, Clegg DJ, Seeley RJ, Woods SC (2002) The catabolic action of insulin in the brain is mediated by melanocortins. *J Neurosci* 22:9048–52.

Berridge MJ, Bootman MD, Roderick HL (2003) Calcium signalling: dynamics, homeostasis and remodelling. *Nat Rev Mol Cell Biol* 4:517–29.

Bouret SG, Draper SJ, Simerly RB (2004) Formation of projection pathways from the arcuate nucleus of the hypothalamus to hypothalamic regions implicated in the neural control of feeding behavior in mice. *The Journal of Neuroscience* 24:2797–2805.

Brüning JC, Gautam D, Burks DJ, Gillette J, Schubert M, Orban PC, Klein R, Krone W, Müller-Wieland D, Kahn CR (2000) Role of brain insulin receptor in control of body weight and reproduction. *Science* 289:2122–5.

Bu J, Sathyendra V, Nagykery N, Geula C (2003) Age-related changes in calbindin-D28k, calretinin, and parvalbumin-immunoreactive neurons in the human cerebral cortex. *Experimental Neurology* 182:220–231.

Campbell LW, Hao SY, Thibault O, Blalock EM, Landfield PW (1996) Aging changes in voltage-gated calcium currents in hippocampal CA1 neurons. *The Journal of Neuroscience* 16:6286–6295.

Campfield LA, Smith FJ, Guisez Y, Devos R, Burn P (1995) Recombinant mouse OB protein: evidence for a peripheral signal linking adiposity and central neural networks. *Science* 269:546–9.

Catterall WA (2000) Structure and regulation of voltage-gated Ca2+ channels. *Annual Review of Cell and Developmental Biology* 16:521–555.

Chambard JM, Chabbert C, Sans A, Desmadryl G (1999) Developmental changes in low and high voltage-activated calcium currents in acutely isolated mouse vestibular neurons. *The Journal of Physiology* 518 (Pt 1):141–149.

Cone RD (2005) Anatomy and regulation of the central melanocortin system. *Nature Neuroscience* 8:571–578.

Connor JA, Stevens CF (1971) Voltage clamp studies of a transient outward membrane current in gastropod neural somata. *The Journal of Physiology* 213:21–30.

Cowley MA, Pronchuk N, Fan W, Dinulescu DM, Colmers WF, Cone RD (1999) Integration of NPY, AGRP, and melanocortin signals in the hypothalamic paraventricular nucleus: evidence of a cellular basis for the adipostat. *Neuron* 24:155–163.

Cowley MA, Smart JL, Rubinstein M, Cerdan MG, Diano S, Horvath TL, Cone RD, Low MJ (2001) Leptin activates anorexigenic POMC neurons through a neural network in the arcuate nucleus. *Nature* 411:480–4.

Dodt HU, Zieglgänsberger W (1990) Visualizing unstained neurons in living brain slices by infrared DIC-videomicroscopy. *Brain Research* 537:333–6.

Dunn-Meynell AA, Rawson NE, Levin BE (1998) Distribution and phenotype of neurons containing the ATP-sensitive K+ channel in rat brain. *Brain Research* 814:41–54.

Efron B (1979) Bootstrap methods: another look at the jackknife. *The Annals of Statistics* 7:1–26.

Elias CF, Aschkenasi C, Lee C, Kelly J, Ahima RS, Bjorbaek C, Flier JS, Saper CB, Elmquist JK (1999) Leptin differentially regulates NPY and POMC neurons projecting to the lateral hypothalamic area. *Neuron* 23:775–86.

Elmquist JK, Ahima RS, Maratos-Flier E, Flier JS, Saper CB (1997) Leptin Activates Neurons in Ventrobasal Hypothalamus and Brainstem. *Endocrinology* 138:839–842.

Elsen FP, Ramirez JM (1998) Calcium currents of rhythmic neurons recorded in the isolated respiratory network of neonatal mice. *The Journal of Neuroscience* 18:10652–10662.

Ema M, Morita M, Ikawa S, Tanaka M, Matsuda Y, Gotoh O, Saijoh Y, Fujii H, Hamada H, Kikuchi Y, Fujii-Kuriyama Y (1996) Two new members of the murine Sim gene family are transcriptional repressors and show different expression patterns during mouse embryogenesis. *Molecular and Cellular Biology* 16:5865–5875.

Ernst MB, Wunderlich CM, Hess S, Paehler M, Mesaros A, Koralov SB, Kleinridders A, Husch A, Münzberg H, Hampel B, Alber J, Kloppenburg P, Brüning JC, Wunderlich FT (2009) Enhanced Stat3 activation in POMC neurons provokes negative feedback inhibition of leptin and insulin signaling in obesity. *The Journal of Neuroscience* 29:11582–11593.

Ertel EA, Campbell KP, Harpold MM, Hofmann F, Mori Y, Perez-Reyes E, Schwartz A, Snutch TP, Tanabe T, Birnbaumer L, Tsien RW, Catterall WA (2000) Nomenclature of voltage-gated calcium channels. *Neuron* 25:533–535.

Faivre L, Cormier-Daire V, Lapierre JM, Colleaux L, Jacquemont S, Geneviéve D, Saunier P, Munnich A, Turleau C, Romana S, Prieur M, Blois MCD, Vekemans M (2002) Deletion of the SIM1 gene (6q16.2) in a patient with a Prader-Willi-like phenotype. *Journal of Medical Genetics* 39:594–596.

Fan W, Boston BA, Kesterson RA, Hruby VJ, Cone RD (1997) Role of melanocortinergic neurons in feeding and the agouti obesity syndrome. *Nature* 385:165–8.

Fierro L, DiPolo R, Llano I (1998) Intracellular calcium clearance in Purkinje cell somata from rat cerebellar slices. *The Journal of Physiology* 510 (Pt 2):499–512.

Fierro L, Llano I (1996) High endogenous calcium buffering in Purkinje cells from rat cerebellar slices. *The Journal of Physiology* 496 (Pt 3):617–625.

Finkelstein EA, Trogdon JG, Cohen JW, Dietz W (2009) Annual Medical Spending Attributable To Obesity: Payer- And Service-Specific Estimates. *Health Aff* p. hlthaff.28.5.w822.

Foehring RC, Zhang XF, Lee JCF, Callaway JC (2009) Endogenous calcium buffering capacity of substantia nigral dopamine neurons. *Journal of Neurophysiology* 102:2326–2333.

Foll CL, Irani BG, Magnan C, Dunn-Meynell AA, Levin BE (2009) Characteristics and mechanisms of hypothalamic neuronal fatty acid sensing. *American Journal of Physiology. Regulatory, Integrative and Comparative Physiology* 297:R655–664.

Frederich RC, Hamann A, Anderson S, Löllmann B, Lowell BB, Flier JS (1995) Leptin levels reflect body lipid content in mice: evidence for diet-induced resistance to leptin action. *Nature Medicine* 1:1311–1314.

Green PJ, Silverman BW (2000) *Nonparametric regression and generalized linear models: a roughness penalty approach* Monographs on statistics and applied probability ; 58. Chapman & Hall, Boca Raton [u.a.], 1. CRC press repr edition.

Gribble FM, Ashfield R, Ammälä C, Ashcroft FM (1997) Properties of cloned ATP-sensitive K+ currents expressed in Xenopus oocytes. *The Journal of Physiology* 498 (Pt 1):87–98.

Grynkiewicz G, Poenie M, Tsien RY (1985) A new generation of Ca2+ indicators with greatly improved fluorescence properties. *J Biol Chem* 260:3440–50.

Gu H, Marth J, Orban P, Mossmann H, Rajewsky K (1994) Deletion of a DNA polymerase beta gene segment in T cells using cell type-specific gene targeting. *Science* 265:103–106.

Guillod-Maximin E, Roy AF, Vacher CM, Aubourg A, Bailleux V, Lorsignol A, Pénicaud L, Parquet M, Taouis M (2009) Adiponectin receptors are expressed in hypothalamus and colocalized with proopiomelanocortin and neuropeptide Y in rodent arcuate neurons. *The Journal of Endocrinology* 200:93–105.

Hahn TM, Breininger JF, Baskin DG, Schwartz MW (1998) Coexpression of Agrp and NPY in fasting-activated hypothalamic neurons. *Nat Neurosci* 1:271–2.

Halaas JL, Gajiwala KS, Maffei M, Cohen SL, Chait BT, Rabinowitz D, Lallone RL, Burley SK, Friedman JM (1995) Weight-reducing effects of the plasma protein encoded by the obese gene. *Science* 269:543–6.

Hamill OP, Marty A, Neher E, Sakmann B, Sigworth FJ (1981) Improved patch-clamp techniques for high-resolution current recording from cells and cell-free membrane patches. *Pflügers Arch* 391:85–100.

Harrison SM, Bers DM (1987) The effect of temperature and ionic strength on the apparent Ca-affinity of EGTA and the analogous Ca-chelators BAPTA and dibromo-BAPTA. *Biochimica Et Biophysica Acta* 925:133–143.

Harrison SM, Bers DM (1989) Correction of proton and Ca association constants of EGTA for temperature and ionic strength. *The American Journal of Physiology* 256:C1250–1256.

Hauner H (2005) Secretory factors from human adipose tissue and their functional role. *The Proceedings of the Nutrition Society* 64:163–169.

Heidel E, Pflüger H (2006) Ion currents and spiking properties of identified subtypes of locust octopaminergic dorsal unpaired median neurons. *The European Journal of Neuroscience* 23:1189–1206.

Helmchen F, Borst JG, Sakmann B (1997) Calcium dynamics associated with a single action potential in a CNS presynaptic terminal. *Biophysical Journal* 72:1458–1471.

Helmchen F, Imoto K, Sakmann B (1996) Ca2+ buffering and action potential-evoked Ca2+ signaling in dendrites of pyramidal neurons. *Biophys J* 70:1069–81.

Helmchen F, Tank DW (2005) A Single-Compartment Model of Calcium Dynamics in Nerve Terminals and Dendrites In *R. Yuste and A. Konnerth (Eds.), Imaging in Neuroscience and Development: A Laboratory Manual*, pp. 265–275. Cold Spring Harbor Laboratory Press, Coldspring Harbor, New York.

Hemond P, Jaffe DB (2005) Caloric restriction prevents aging-associated changes in spike-mediated Ca2+ accumulation and the slow afterhyperpolarization in hippocampal CA1 pyramidal neurons. *Neuroscience* 135:413–420.

Hentges ST, Otero-Corchon V, Pennock RL, King CM, Low MJ (2009) Proopiomelanocortin Expression in both GABA and Glutamate Neurons. *J. Neurosci.* 29:13684–13690.

Hetherington AW, Ranson SW (1940) Hypothalamic lesions and adipostiy in the rat. *Anat. Rec.* 78:149 – 172.

Hille B (2001) *Ion Channels of Excitable Membranes*, Vol. 3 Sinauer Associates, Inc., Sunderland, Massachusetts USA.

Hoffman NW, Tasker JG, Dudek FE (1991) Immunohistochemical differentiation of electrophysiologically defined neuronal populations in the region of the rat hypothalamic paraventricular nucleus. *The Journal of Comparative Neurology* 307:405–416.

Holder JL, Butte NF, Zinn AR (2000) Profound obesity associated with a balanced translocation that disrupts the SIM1 gene. *Human Molecular Genetics* 9:101–108.

Horikawa K, Armstrong WE (1988) A versatile means of intracellular labeling: injection of biocytin and its detection with avidin conjugates. *J Neurosci Methods* 25:1–11.

Horn R, Marty A (1988) Muscarinic activation of ionic currents measured by a new whole-cell recording method. *The Journal of General Physiology* 92:145–159.

Huxley RR, Ansary-Moghaddam A, Clifton P, Czernichow S, Parr CL, Woodward M (2009) The impact of dietary and lifestyle risk factors on risk of colorectal cancer: A quantitative overview of the epidemiological evidence. *International Journal of Cancer. Journal International Du Cancer* .

Kammermeier PJ, Jones SW (1997) High-voltage-activated calcium currents in neurons acutely isolated from the ventrobasal nucleus of the rat thalamus. *Journal of Neurophysiology* 77:465–475.

Kay JW, Steven RJ, McGuigan JA, Elder HY (2008) Automatic determination of ligand purity and apparent dissociation constant (K(app)) in Ca(2+)/Mg(2+) buffer solutions and the K(app) for Ca(2+)/Mg(2+) anion binding in physiological solutions from Ca(2+)/Mg(2+)-macroelectrode measurements. *Comput Biol Med* 38:101–10.

Khachaturian ZS (1987) Hypothesis on the regulation of cytosol calcium concentration and the aging brain. *Neurobiology of Aging* 8:345–346.

Kirischuk S, Pronchuk N, Verkhratsky A (1992) Measurements of intracellular calcium in sensory neurons of adult and old rats. *Neuroscience* 50:947–951.

Kirischuk S, Verkhratsky A (1996) Calcium homeostasis in aged neurones. *Life Sciences* 59:451–459.

Kloppenburg P, Zipfel WR, Webb WW, Harris-Warrick RM (2000) Highly localized Ca(2+) accumulation revealed by multiphoton microscopy in an identified motoneuron and its modulation by dopamine. *J Neurosci* 20:2523–33.

Kloppenburg P, Zipfel WR, Webb WW, Harris-Warrick RM (2007) Heterogeneous Effects of Dopamine on Highly Localized, Voltage-Induced Ca2+ Accumulation in Identified Motoneurons. *J Neurophysiol* 98:2910–2917.

Kostyuk P, Pronchuk N, Savchenko A, Verkhratsky A (1993) Calcium currents in aged rat dorsal root ganglion neurones. *The Journal of Physiology* 461:467–483.

Landfield PW (1987) 'Increased calcium-current' hypothesis of brain aging. *Neurobiology of Aging* 8:346–347.

Lee D, Obukhov AG, Shen Q, Liu Y, Dhawan P, Nowycky MC, Christakos S (2006) Calbindin-D28k decreases L-type calcium channel activity and modulates intracellular calcium homeostasis in response to K+ depolarization in a rat beta cell line RINr1046-38. *Cell Calcium* 39:475–485.

Lee SH, Rosenmund C, Schwaller B, Neher E (2000) Differences in Ca2+ buffering properties between excitatory and inhibitory hippocampal neurons from the rat. *J Physiol* 525:405–18.

Lee S, Chen Y, Chen Y, Chang S, Tai C, Wongcharoen W, Yeh H, Lin C, Chen S (2007) Tumor necrosis factor-alpha alters calcium handling and increases arrhythmogenesis of pulmonary vein cardiomyocytes. *Life Sciences* 80:1806–1815.

Liao C, Lien C (2009) Estimating intracellular Ca2+ concentrations and buffering in a dendritic inhibitory hippocampal interneuron. *Neuroscience* 164:1701–1711.

Lin S, Thomas TC, Storlien LH, Huang XF (2000) Development of high fat diet-induced obesity and leptin resistance in C57Bl/6J mice. *International Journal of Obesity and Related Metabolic Disorders: Journal of the International Association for the Study of Obesity* 24:639–646.

Lincoln DW, Wakerley JB (1974) Electrophysiological evidence for the activation of supraoptic neurones during the release of oxytocin. *The Journal of Physiology* 242:533–554.

Lips MB, Keller BU (1998) Endogenous calcium buffering in motoneurones of the nucleus hypoglossus from mouse. *The Journal of Physiology* 511 (Pt 1):105–117.

Liss B, Roeper J (2004) Correlating function and gene expression of individual basal ganglia neurons. *Trends in Neurosciences* 27:475–481.

Lüthi D, Spichiger U, Forster I, McGuigan JA (1997) Calibration of Mg(2+)-selective macroelectrodes down to 1 mumol l-1 in intracellular and Ca(2+)-containing extracellular solutions. *Exp Physiol* 82:453–67.

Lu S, Gold MS (2008) Inflammation-induced increase in evoked calcium transients in subpopulations of rat dorsal root ganglion neurons. *Neuroscience* 153:279–288.

Luchsinger JA, Mayeux R (2007) Adiposity and Alzheimer's disease. *Current Alzheimer Research* 4:127–34.

Luther JA, Daftary SS, Boudaba C, Gould GC, Halmos KC, Tasker JG (2002) Neurosecretory and non-neurosecretory parvocellular neurones of the hypothalamic paraventricular nucleus express distinct electrophysiological properties. *Journal of Neuroendocrinology* 14:929–932.

Luther JA, Halmos KC, Tasker JG (2000) A slow transient potassium current expressed in a subset of neurosecretory neurons of the hypothalamic paraventricular nucleus. *Journal of Neurophysiology* 84:1814–1825.

Luther JA, Tasker JG (2000) Voltage-gated currents distinguish parvocellular from magnocellular neurones in the rat hypothalamic paraventricular nucleus. *The Journal of Physiology* 523 Pt 1:193–209.

Malpas SC, Coote JH (1994) Role of vasopressin in sympathetic response to paraventricular nucleus stimulation in anesthetized rats. *Am J Physiol Regul Integr Comp Physiol* 266:R228–236.

Marambaud P, Dreses-Werringloer U, Vingtdeux V (2009) Calcium signaling in neurodegeneration. *Molecular Neurodegeneration* 4:20.

Marion J, Yang C, Caqueret A, Boucher F, Michaud JL (2005) Sim1 and Sim2 are required for the correct targeting of mammillary body axons. *Development (Cambridge, England)* 132:5527–5537.

Martella G, Spadoni F, Sciamanna G, Tassone A, Bernardi G, Pisani A, Bonsi P (2008) Age-related functional changes of high-voltage-activated calcium channels in different neuronal subtypes of mouse striatum. *Neuroscience* 152:469–476.

Mattson MP, Chan SL (2003) Neuronal and glial calcium signaling in Alzheimer's disease. *Cell Calcium* 34:385–397.

McGuigan JA, Kay JW, Elder HY (2006) Critical review of the methods used to measure the apparent dissociation constant and ligand purity in Ca2+ and Mg2+ buffer solutions. *Prog Biophys Mol Biol* 92:333–70.

McGuigan JA, Lüthi D, Buri A (1991) Calcium buffer solutions and how to make them: a do it yourself guide. *Can J Physiol Pharmacol* 69:1733–49.

McGuigan JAS, Kay JW, Elder HY, Lüthi D (2007) Comparison between measured and calculated ionised concentrations in Mg2+ /ATP, Mg2+ /EDTA and Ca2+ /EGTA buffers; influence of changes in temperature, pH and pipetting errors on the ionised concentrations. *Magnesium Research: Official Organ of the International Society for the Development of Research on Magnesium* 20:72–81.

Mehrke G, Zong XG, Flockerzi V, Hofmann F (1994) The Ca(++)-channel blocker Ro 40-5967 blocks differently T-type and L-type Ca++ channels. *The Journal of Pharmacology and Experimental Therapeutics* 271:1483–1488.

Meister B, Gömüç B, Suarez E, Ishii Y, Dürr K, Gillberg L (2006) Hypothalamic proopiomelanocortin (POMC) neurons have a cholinergic phenotype. *The European Journal of Neuroscience* 24:2731–2740.

Melnick I, Pronchuk N, Cowley MA, Grove KL, Colmers WF (2007) Developmental switch in neuropeptide Y and melanocortin effects in the paraventricular nucleus of the hypothalamus. *Neuron* 56:1103–1115.

Mercer JG, Hoggard N, Williams LM, Lawrence CB, Hannah LT, Trayhurn P (1996) Localization of leptin receptor mRNA and the long form splice variant (Ob-Rb) in mouse hypothalamus and adjacent brain regions by in situ hybridization. *FEBS Lett* 387:113–6.

Meuth SG, Kanyshkova T, Landgraf P, Pape H, Budde T (2005) Influence of Ca^{2+}-binding proteins and the cytoskeleton on Ca^{2+}-dependent inactivation of high-voltage activated Ca^{2+} currents in thalamocortical relay neurons. *Pflügers Archiv: European Journal of Physiology* 450:111–122.

Miki T, Liss B, Minami K, Shiuchi T, Saraya A, Kashima Y, Horiuchi M, Ashcroft F, Minokoshi Y, Roeper J, Seino S (2001) ATP-sensitive K^+ channels in the hypothalamus are essential for the maintenance of glucose homeostasis. *Nature Neuroscience* 4:507–512.

Miller NE, Bailey CJ, Stevenson JAF (1950) Decreased "hunger" but increased food intake resulting from hypothalamic lesions. *Science (New York, N.Y.)* 112:256–259.

Moore ED, Becker PL, Fogarty KE, Williams DA, Fay FS (1990) Ca^{2+} imaging in single living cells: theoretical and practical issues. *Cell Calcium* 11:157–79.

Morton GJ, Cummings DE, Baskin DG, Barsh GS, Schwartz MW (2006) Central nervous system control of food intake and body weight. *Nature* 443:289–95.

Motagally MA, Lukewich MK, Chisholm SP, Neshat S, Lomax AE (2009) Tumour necrosis factor alpha activates nuclear factor kappaB signalling to reduce N-type voltage-gated Ca^{2+} current in postganglionic sympathetic neurons. *The Journal of Physiology* 587:2623–2634.

Murchison D, Griffith WH (1995) Low-voltage activated calcium currents increase in basal forebrain neurons from aged rats. *Journal of Neurophysiology* 74:876–887.

Murchison D, Griffith WH (1996) High-voltage-activated calcium currents in basal forebrain neurons during aging. *Journal of Neurophysiology* 76:158–174.

Murchison D, Griffith WH (1998) Increased calcium buffering in basal forebrain neurons during aging. *Journal of Neurophysiology* 80:350–364.

Murchison D, Griffith WH (2007) Calcium buffering systems and calcium signaling in aged rat basal forebrain neurons. *Aging Cell* 6:297–305.

Must A, Spadano J, Coakley EH, Field AE, Colditz G, Dietz WH (1999) The disease burden associated with overweight and obesity. *JAMA: The Journal of the American Medical Association* 282:1523–9.

Nambu JR, Lewis JO, Wharton KA, Crews ST (1991) The Drosophila single-minded gene encodes a helix-loop-helix protein that acts as a master regulator of CNS midline development. *Cell* 67:1157–1167.

Neher E (1989) Combined fura-2 and patch clamp measurements in rat peritoneal mast cells In Sellin LC, Libelius R, Thesleff S, editors, *Neuromuscular Junction*, Vol. 5, pp. 65–76. Elsevier Science Publishers, Amsterdam.

Neher E (1992) Correction for liquid junction potentials in patch clamp experiments. *Methods in Enzymology* 207:123–131.

Neher E, Augustine GJ (1992) Calcium gradients and buffers in bovine chromaffin cells. *J Physiol* 450:273–301.

Nernst W (1888) Zur Kinetik der in Lösung befindlichen Körper: Theorie der Diffusion. *Z. Phys. Chem.* 2:613–637.

Osmond JM, Mintz JD, Dalton B, Stepp DW (2009) Obesity increases blood pressure, cerebral vascular remodeling, and severity of stroke in the Zucker rat. *Hypertension* 53:381–6.

References

Patton C, Thompson S, Epel D (2004) Some precautions in using chelators to buffer metals in biological solutions. *Cell Calcium* 35:427–31.

Paxinos G, Franklin KBJ (1997) *The mouse brain in stereotaxic coordinates.* Academic Press, San Diego, second edition.

Pelleymounter MA, Cullen MJ, Baker MB, Hecht R, Winters D, Boone T, Collins F (1995) Effects of the obese gene product on body weight regulation in ob/ob mice. *Science* 269:540–3.

Perez-Reyes E (2003) Molecular physiology of low-voltage-activated t-type calcium channels. *Physiological Reviews* 83:117–161.

Pippow A, Husch A, Pouzat C, Kloppenburg P (2009) Differences of Ca(2+) handling properties in identified central olfactory neurons of the antennal lobe. *Cell Calcium* 46:87–98.

Poenie M (1990) Alteration of intracellular Fura-2 fluorescence by viscosity: a simple correction. *Cell Calcium* 11:85–91.

Pottorf WJ, Duckles SP, Buchholz JN (2002) Aging and calcium buffering in adrenergic neurons. *Autonomic Neuroscience: Basic & Clinical* 96:2–7.

Pusch M, Neher E (1988) Rates of diffusional exchange between small cells and a measuring patch pipette. *Pflügers Arch* 411:204–11.

Qiu Z, Parsons K, Gruol D (1995) Interleukin-6 selectively enhances the intracellular calcium response to NMDA in developing CNS neurons. *J. Neurosci.* 15:6688–6699.

Randall A, Tsien RW (1995) Pharmacological dissection of multiple types of Ca2+ channel currents in rat cerebellar granule neurons. *The Journal of Neuroscience* 15:2995–3012.

Rondinone CM (2006) Adipocyte-derived hormones, cytokines, and mediators. *Endocrine* 29:81–90.

Rush ME, Rinzel J (1995) The potassium A-current, low firing rates and rebound excitation in Hodgkin-Huxley models. *Bulletin of Mathematical Biology* 57:899–929.

Sakura H, Ammälä C, Smith PA, Gribble FM, Ashcroft FM (1995) Cloning and functional expression of the cDNA encoding a novel ATP-sensitive potassium channel subunit expressed in pancreatic beta-cells, brain, heart and skeletal muscle. *FEBS Letters* 377:338–344.

Saltiel AR, Kahn CR (2001) Insulin signalling and the regulation of glucose and lipid metabolism. *Nature* 414:799–806.

Sauer B, Henderson N (1988) Site-specific DNA recombination in mammalian cells by the Cre recombinase of bacteriophage P1. *Proceedings of the National Academy of Sciences of the United States of America* 85:5166–5170.

Sawchenko PE, Swanson LW (1982) Immunohistochemical identification of neurons in the paraventricular nucleus of the hypothalamus that project to the medulla or to the spinal cord in the rat. *The Journal of Comparative Neurology* 205:260–272.

Schwartz MW, Porte D (2005) Diabetes, Obesity, and the Brain. *Science* 307:375–379.

Shirasaka T, Yoshimura Y, Qiu D, Takasaki M (2004) The effects of propofol on hypothalamic paraventricular nucleus neurons in the rat. *Anesthesia and Analgesia* 98:1017–1023, table of contents.

Stern JE (2001) Electrophysiological and morphological properties of pre-autonomic neurones in the rat hypothalamic paraventricular nucleus. *The Journal of Physiology* 537:161–177.

Stevenson JAF (1970) Neural control of food and water intake In *The Hypothalamus*, pp. 524 – 612. C. C. Thomas, Springfield, IL.

Stocker SD, Cunningham JT, Toney GM (2004) Water deprivation increases Fos immunoreactivity in PVN autonomic neurons with projections to the spinal cord and rostral ventrolateral medulla. *American Journal of Physiology. Regulatory, Integrative and Comparative Physiology* 287:R1172–1183.

Stuenkel EL (1994) Regulation of intracellular calcium and calcium buffering properties of rat isolated neurohypophysial nerve endings. *The Journal of Physiology* 481 (Pt 2):251–271.

Summerlee AJ, Lincoln DW (1981) Electrophysiological recordings from oxytocinergic neurones during suckling in the unanaesthetized lactating rat. *The Journal of Endocrinology* 90:255–265.

Surwit RS, Kuhn CM, Cochrane C, McCubbin JA, Feinglos MN (1988) Diet-induced type II diabetes in C57BL/6J mice. *Diabetes* 37:1163–7.

Swanson LW (2000) Cerebral hemisphere regulation of motivated behavior. *Brain Research* 886:113–164.

Swanson LW, Sawchenko PE (1983) Hypothalamic integration: organization of the paraventricular and supraoptic nuclei. *Annual Review of Neuroscience* 6:269–324.

Swanson LW, Sawchenko PE, Wiegand SJ, Price JL (1980) Separate neurons in the paraventricular nucleus project to the median eminence and to the medulla or spinal cord. *Brain Research* 198:190–195.

Tanaka K, Shirakawa H, Okada K, Konno M, Nakagawa T, Serikawa T, Kaneko S (2007) Increased Ca2+ channel currents in cerebellar Purkinje cells of the ataxic groggy rat. *Neuroscience Letters* 426:75–80.

Tang CM, Presser F, Morad M (1988) Amiloride selectively blocks the low threshold (T) calcium channel. *Science (New York, N.Y.)* 240:213–215.

Tank DW, Regehr WG, Delaney KR (1995) A quantitative analysis of presynaptic calcium dynamics that contribute to short-term enhancement. *J Neurosci* 15:7940–52.

Tartaglia LA, Dembski M, Weng X, Deng N, Culpepper J, Devos R, Richards GJ, Campfield LA, Clark FT, Deeds J, Muir C, Sanker S, Moriarty A, Moore KJ, Smutko JS, Mays GG, Wool EA, Monroe CA, Tepper RI (1995) Identification and expression cloning of a leptin receptor, OB-R. *Cell* 83:1263–71.

Tasker JG, Dudek FE (1991) Electrophysiological properties of neurones in the region of the paraventricular nucleus in slices of rat hypothalamus. *The Journal of Physiology* 434:271–293.

Tasker JG, Dudek FE (1993) Local inhibitory synaptic inputs to neurones of the paraventricular nucleus in slices of rat hypothalamus. *The Journal of Physiology* 469:179–192.

Tatsumi H, Katayama Y (1993) Regulation of the intracellular free calcium concentration in acutely dissociated neurones from rat nucleus basalis. *J Physiol* 464:165–81.

Thibault O, Landfield PW (1996) Increase in single L-type calcium channels in hippocampal neurons during aging. *Science (New York, N.Y.)* 272:1017–1020.

Thompson D, Edelsberg J, Colditz GA, Bird AP, Oster G (1999) Lifetime health and economic consequences of obesity. *Archives of Internal Medicine* 159:2177–83.

Tilg H, Moschen AR (2006) Adipocytokines: mediators linking adipose tissue, inflammation and immunity. *Nature Reviews. Immunology* 6:772–783.

Toescu EC, Verkhratsky A (2007) The importance of being subtle: small changes in calcium homeostasis control cognitive decline in normal aging. *Aging Cell* 6:267–273.

Tonkikh A, Janus C, El-Beheiry H, Pennefather PS, Samoilova M, McDonald P, Ouanounou A, Carlen PL (2006) Calcium chelation improves spatial learning and synaptic plasticity in aged rats. *Experimental Neurology* 197:291–300.

Trayhurn P, Beattie JH (2001) Physiological role of adipose tissue: white adipose tissue as an endocrine and secretory organ. *The Proceedings of the Nutrition Society* 60:329–339.

Triggle DJ (2006) L-type calcium channels. *Current Pharmaceutical Design* 12:443–457.

Vanselow BK, Keller BU (2000) Calcium dynamics and buffering in oculomotor neurones from mouse that are particularly resistant during amyotrophic lateral sclerosis (ALS)-related motoneurone disease. *J Physiol* 525 Pt 2:433–45.

Verkhratsky A, Toescu EC (2003) Endoplasmic reticulum Ca(2+) homeostasis and neuronal death. *Journal of Cellular and Molecular Medicine* 7:351–361.

Vrang N, Larsen PJ, Clausen JT, Kristensen P (1999) Neurochemical characterization of hypothalamic cocaine- amphetamine-regulated transcript neurons. *The Journal of Neuroscience* 19:RC5.

Wanaverbecq N, Marsh SJ, Al-Qatari M, Brown DA (2003) The plasma membrane calcium-ATPase as a major mechanism for intracellular calcium regulation in neurones from the rat superior cervical ganglion. *The Journal of Physiology* 550:83–101.

Wang J, Wang F, Yang M, Yu D, Wu W, Liu J, Ma L, Cai F, Chen J (2008) Leptin regulated calcium channels of neuropeptide Y and proopiomelanocortin neurons by activation of different signal pathways. *Neuroscience* 156:89–98.

West DB, Boozer CN, Moody DL, Atkinson RL (1992) Dietary obesity in nine inbred mouse strains. *The American Journal of Physiology* 262:R1025–32.

White JA, Sekar NS, Kay AR (1995) Errors in persistent inward currents generated by space-clamp errors: a modeling study. *J Neurophysiol* 73:2369–2377.

Wicher D, Penzlin H (1997) Ca2+ currents in central insect neurons: electrophysiological and pharmacological properties. *Journal of Neurophysiology* 77:186–199.

World Health Organization (2009) *Global health risks. Mortality and burden of disease attributable to selected major risks.* WHO Press, Geneva, Switzerland.

Yaffe K (2007) Metabolic syndrome and cognitive disorders: is the sum greater than its parts? *Alzheimer Disease and Associated Disorders* 21:167–71.

Yarom Y, Sugimori M, Llinás R (1985) Ionic currents and firing patterns of mammalian vagal motoneurons in vitro. *Neuroscience* 16:719–737.

Ye JH, Zhang J, Xiao C, Kong J (2006) Patch-clamp studies in the CNS illustrate a simple new method for obtaining viable neurons in rat brain slices: glycerol replacement of NaCl protects CNS neurons. *Journal of Neuroscience Methods* 158:251–9.

Zhang W, Star B, Rajapaksha WR, Fisher TE (2007) Dehydration increases L-type Ca(2+) current in rat supraoptic neurons. *The Journal of Physiology* 580:181–193.

Zhang Y, Proenca R, Maffei M, Barone M, Leopold L, Friedman JM (1994) Positional cloning of the mouse obese gene and its human homologue. *Nature* 372:425–32.

Zhou Z, Neher E (1993) Mobile and immobile calcium buffers in bovine adrenal chromaffin cells. *The Journal of Physiology* 469:245–273.

Acknowledgements

A PhD thesis is rarely the work of one single person alone. I would like to acknowledge the following people for helping me one way or the other:

Prof. Dr. Peter Kloppenburg for giving me the opportunity to write this thesis, the excellent supervision, support and all the fruitful discussions. Prof. Dr. Jens Brüning for his support in the collaboration and during this thesis. In addition, I would like to thank both Profs. Kloppenburg and Brüning for their critically important input and support for the successful grant application for the Boehringer Ingelheim Fonds PhD scholarship.

I would like to thank Prof. Dr. Ansgar Büschges for being the second referee to this thesis and for the many enjoyable conversations.

I am also greatly indebted to the Boehringer Ingelheim Fonds especially to
Dr. Herrmann Fröhlich, Monika Beutelspacher, Dr. Claudia Walther and Dr. Sabine Achten, not only for their trust, encouragement and the opportunity to meet and communicate with fellow stipend holders but also for the generous financial support during this thesis.

Helmut Wratil for his dedicated and truly excellent technical assistance and advice.

Sincere thanks go to Simon Heß and Dr. Andreas Pippow. Without your help this thesis in its present form would not have been possible. But also for being cool dudes.

Nora Redemann for the enjoyable collaboration on the SIM1 project and the steady supply of mice. Lars Paeger for his valuable help during the SIM1 project.

Dr. Eva Rother, Tim Klöckener and Bengt Belgardt for their excellent mouse supply chain and collaborative effort.

Eugenio Oliveira for the enjoyable working experience and for imparting a bit of the brazilian take on things on me. Ordem e progresso!

Florian Leiß for the enjoyable collaboration and good times in Munich and Cologne.

The Kloppenburg-lab: Sabine, Debora, Heike, Andreas, Cathleen for generating a pleasant working environment and for enforcing a strict dining schedule.

Prof. Dr. Ron Harris-Warrick not only for the good times and great conversations but also for being a mindblowingly cool guy.

Dr. Bruce Johnson for his input on this thesis and for the enjoyable conversations.

My senior biology teacher Annemarie Wehner-Eckl, to whom this thesis is dedicated, for encouraging and enduring my curiosity and enthusiasm and for being rather blunt about it, too..

Dani, Tina, Steffi, Claus, Maize, Christian, Alex, Jens, Leo, Robert, Florian, Henry, Jan for being great friends, for lending me an open ear when dearly needed, for the laughs we had and the ones that are still to be had.

Matthias Könn, Claus Bender, Alexander Krupp and Jens Frauenfeld for their eagle-eye proofreading.

The 1.FC Köln for sometimes being the new standard of godawful sucking-ness and Acki, Rams and Jens for enduring it with me.

My siblings and their spouses, Anke & Kay, Tim & Petra and Jan & Katrin for their support and their limitless compassion during the highs and lows of this thesis.

My nieces and nephews, Smilla, Florian, Leo, Nick, Max and Karla for constantly correcting my perspective on the more fundamental concerns in life, for their inspiring curiosity and their hilarious ability to amaze.

And ultimately, my loving parents Hans & Katrin who never faltered in their belief in me and who have always picked me up, encouraged and supported me in every way possible. Also, for proofreading and comments on the manuscript. :) Thank You!

Finally, Otto Loewi for saying: "Ja, Kalzium, das ist alles.." and being dead wrong...

Erklärung

Ich versichere, dass ich die von mir vorgelegte Dissertation selbständig angefertigt, die benutzten Quellen und Hilfsmittel vollständig angegeben und die Stellen der Arbeit - einschließlich Tabellen, Karten und Abbildungen -, die anderen Werken im Wortlaut oder dem Sinn nach entnommen sind, in jedem Einzelfall als Entlehnung kenntlich gemacht habe; dass diese Dissertation noch keiner anderen Fakultät oder Universität zur Prüfung vorgelegen hat; dass sie - abgesehen von unten angegebenen Teilpublikationen - noch nicht veröffentlicht worden ist sowie, dass ich eine solche Veröffentlichung vor Abschluss des Promotionsverfahrens nicht vornehmen werde. Die Bestimmungen dieser Promotionsordnung sind mir bekannt. Die von mir vorgelegte Dissertation ist von Prof. Dr. Peter Kloppenburg betreut worden.

Köln, den 4. Dezember 2009

Teilpublikationen

Articles

Ernst, MB., Wunderlich, CM., Hess, S., **Paehler, M.**, Mesaros, A., Koralov, SB. Kleinridders, A., Husch, A., Münzberg, H., Hampel, B., Alber, J., Kloppenburg, P., Brüning, JC., Wunderlich, TF. (2009). Enhanced Stat3-activation in POMC-neurons provokes negative feedback inhibition of leptin- and insulin-signaling in obesity. *J Neurosci 29(37):11582-93.*

Husch, A., **Paehler, M.**, Fusca, D., Paeger, L. and Kloppenburg, P. (2009). Distinct electrophysiological properties in sub-types of non-spiking olfactory local interneurons correlate with their cell type specific Ca^{2+} current profiles. *J Neurophysiol 102(5):2834-45.*

Husch, A., **Paehler, M.**, Fusca, D., Paeger, L. and Kloppenburg, P. (2008). Calcium current diversity in physiologically different local interneuron types of the antennal lobe. *J Neurosci 29(3):716-26.*

Posters and Abstracts

Hess, S., Husch, A., **Paehler, M.**, Pippow, A., Wratil,H., Belgardt, BF., Kloeckener, T., Rother, E., Brüning, JC., and Kloppenburg, P. (2009). Functional parameters of identified neurons in the arcuate nucleus of the hypothalamus. *Annual Meeting of the Society for Neuroscience (SfN). Abstract, Chicago, IL*

Husch, A., Fusca, D., **Paehler, M.**, Wratil,H., and Kloppenburg, P. (2009). Distinct subtypes of local interneurons in the olfactory system of periplaneta americana. *Annual Meeting of the Society for Neuroscience (SfN). Abstract, Chicago, IL*

Pippow, A., Demmer, H., Fusca, D., Hess, S., Husch, A., **Paehler, M.**, Wratil,H., Pouzat,C. and Kloppenburg, P. (2008). Distinct calcium handling properties of identified insect olfactory interneurons. *Annual Meeting of the Society for Neuroscience (SfN). Program No. 363.5 Abstract, Washington, DC.*

Demmer, H., Hess, S., Husch, A., **Paehler, M.** and Kloppenburg, P. (2007). Physiological and morphological characterisation of interneurons in the insect olfactory pathway. *Annual Meeting of the Society for Neuroscience (SfN). Program No. 277.8 Abstract, San Diego, CA*

Husch A., **Paehler M.**, and Kloppenburg P. (2007). Characterization of local interneurons in the antennal lobe of periplaneta americana. *Proceedings of the 100th Meeting of the German Zoological Society, Abstract, Cologne.*

Husch A., **Paehler M.**, and Kloppenburg P. (2007). Electrophysiological and morphological characterization of spiking and nonspiking local interneurons in the antennal lobe of periplaneta americana. *Proceedings of the 31th Göttingen Neurobiology Conference and the 7th Meeting of the German Neuroscience Society.*

Pippow A., Husch A., **Paehler M.** and Kloppenburg P. (2006). Calcium dynamics in olfactory interneurons in situ. *5th Forum of European Neuroscience (Fens). A074.15*

Husch A., **Paehler M.**, and Kloppenburg P. (2006). Structural and functional properties of olfactory interneurons in the insect antennal lobe. *Neurovisionen: Perspektiven in NRW, Düsseldorf, Germany.*

Curriculum Vitae

Dipl. Biol. Moritz Paehler
Campus address:
Institute of Zoology, Biozentrum, University of Cologne
Cellular and Molecular Neurophysiology
Otto-Fischer Str. 6, D-50674 Cologne, Germany
Email: moritz.paehler@uni-koeln.de
Phone: +49221-470-5828
Date of Birth: 13.06.1980
Place of Birth: Bonn
Citizenship: German

Education

	2006 - 2010	PhD thesis in neurobiology, University of Cologne
	2006	Diplom in biology (Master equivalent) with majors in zoology, botany and biochemistry; title of thesis: a method for functional *in situ* calcium-imaging in olfactory interneurons of *Periplaneta americana*, University of Cologne
	September 2005	Advanced Training in Immunohistochemistry, EURON - European Graduate School of Neuroscience, University of Maastricht, NL
	2003	Vordiplom in biology (Bachelor equivalent), University of Cologne
	2000	Abitur (University entrance exam), Konrad-Adenauer-Gymnasium, Meckenheim

Awards

	March 2007 - September 2009	Boehringer Ingelheim Fonds PhD Scholarship

Professional Experiences

	2005 - 2007	Graduate assistant in the lab of Prof. Dr. Peter Kloppenburg, University of Cologne
	March - May 2005	Internship with the editorial staff of "nano" at the ZDF (Science program, German television)
	2002 - 2005	Student assistant for the visualisation of highly complex ligand molecules using computer aided drawing software for the BRENDA Enzyme Database at the Institute of Biochemistry, University of Cologne

Lebenslauf

Dipl. Biol. Moritz Paehler
Campusadresse:
Zoologisches Institut, Biozentrum, Universität zu Köln
Zelluläre und Molekulare Neurophysiologie
Otto-Fischer Str. 6, D-50674 Köln
Email: moritz.paehler@uni-koeln.de
Telefon: +49221-470-5828
Geboren am: 13.06.1980
Geboren in: Bonn
Staatsangehörigkeit: deutsch

Ausbildung

2006 - 2010	Doktorarbeit in Neurobiologie, Universität zu Köln
2006	Diplom in Biologie in den Fächern Zoologie, Botanik and Biochemie; Titel der Diplomarbeit: A method for functional *in situ* calcium-imaging in olfactory interneurons of *Periplaneta americana*, Universität zu Köln
September 2005	Fortbildung: Advanced Training in Immunohistochemistry, EURON - European Graduate School of Neuroscience, University of Maastricht, NL
2003	Vordiplom in Biologie, Universität zu Köln
2000	Abitur, Konrad-Adenauer-Gymnasium, Meckenheim

Auszeichnungen

März 2007 - September 2009	Boehringer Ingelheim Fonds PhD-Stipendium

Berufliche Tätigkeiten

2005 - 2007	Wissenschaftliche Hilfskraft im Labor von Prof. Dr. Peter Kloppenburg, Universität zu Köln
März - Mai 2005	Praktikum in der Redation nano"beim ZDF
2002 - 2005	Studentische Hilfskraft für die visualisierung hochkomplexer Ligandenmoleküle mit Hilfe Computer gestützer Zeichensoftware für die BRENDA Enzyme Database im Institut für Biochemie, Universität zu Köln

I want morebooks!

Buy your books fast and straightforward online - at one of world's fastest growing online book stores! Environmentally sound due to Print-on-Demand technologies.

Buy your books online at
www.morebooks.shop

Kaufen Sie Ihre Bücher schnell und unkompliziert online – auf einer der am schnellsten wachsenden Buchhandelsplattformen weltweit! Dank Print-On-Demand umwelt- und ressourcenschonend produziert.

Bücher schneller online kaufen
www.morebooks.shop

KS OmniScriptum Publishing
Brivibas gatve 197
LV-1039 Riga, Latvia
Telefax: +371 686 204 55

info@omniscriptum.com
www.omniscriptum.com

Printed by Books on Demand GmbH, Norderstedt / Germany